Springer Series in
CHEMICAL PHYSICS 70

Springer
Berlin
Heidelberg
New York
Hong Kong
London
Milan
Paris
Tokyo

Springer Series in
CHEMICAL PHYSICS

Series Editors: F. P. Schäfer J. P. Toennies W. Zinth

The purpose of this series is to provide comprehensive up-to-date monographs in both well established disciplines and emerging research areas within the broad fields of chemical physics and physical chemistry. The books deal with both fundamental science and applications, and may have either a theoretical or an experimental emphasis. They are aimed primarily at researchers and graduate students in chemical physics and related fields.

63 **Ultrafast Phenomena XI**
Editors: T. Elsaesser, J.G. Fujimoto, D.A. Wiersma, and W. Zinth

64 **Asymptotic Methods in Quantum Mechanics**
Application to Atoms, Molecules and Nuclei
By S.H. Patil and K.T. Tang

65 **Fluorescence Correlation Spectroscopy**
Theory and Applications
Editors: R. Rigler and E.S. Elson

66 **Ultrafast Phenomena XII**
Editors: T. Elsaesser, S. Mukamel, M.M. Murnane, and N.F. Scherer

67 **Single Molecule Spectroscopy**
Nobel Conference Lectures
Editors: R. Rigler, M. Orrit, T. Basché

68 **Nonequilibrium Nondissipative Thermodynamics**
With Application to Low-Pressure Diamond Synthesis
By J.-T. Wang

69 **Selective Spectroscopy of Single Molecules**
By I.S. Osad'ko

70 **Chemistry of Nanomolecular Systems**
Towards the Realization of Molecular Devices
Editors: T. Nakamura, T. Matsumoto, H. Tada, K.-I. Sugiura

71 **Ultrafast PhenomenaXIII**
Editors: D. Miller, M.M. Murnane, N.R. Scherer, and A.M. Weiner

72 **Physical Chemistry of Polymer Rheoloy**
By J. Furukawa

73 **Organometallic Conjugation**
Structures, Reactions and Functions
of d–d and d–π Conjugated Systems
Editors: A. Nakamura, N. Ueyama, and K. Yamaguchi

Series homepage – http://www.springer.de/phys/books/chemical-physics/

For informations about Vols. 1–62 please contact your bookseller

T. Nakamura, T. Matsumoto, H. Tada,
K.-I. Sugiura (Eds.)

Chemistry of Nanomolecular Systems

Towards the Realization of Molecular Devices

With 113 Figures and 3 Tables

Springer

Prof. Takayoshi Nakamura
Research Institute for Electronic Science
Hokkaido University
N12W6, Kita-ku
Sapporo 060-0812, Japan

Prof. Takuya Matsumoto
Institute of Scientific and Industrial Research
Osaka University
8-1 Mihogaoka, Ibaraki
Osaka 567-0047, Japan

Prof. Hirokazu Tada
Institute for Molecular Science
Myodaiji
Okazaki 444-8585, Japan

Prof. Ken-ichi Sugiura
Department of Chemistry
Graduate School of Science
Tokyo Metropolitan University
1-1 Minami-Ohsawa, Hachi-Oji
Tokyo 192-0397, Japan

Series Editors:

Professor F.P. Schäfer
Max-Planck-Institut für Biophysikalische Chemie
37077 Göttingen-Nikolausberg, Germany

Professor J.P. Toennies
Max-Planck-Institut für Strömungsforschung
Bunsenstrasse 10
37073 Göttingen, Germany

Professor W. Zinth
Universität München,
Institut für Medizinische Optik
Öttingerstr. 67
80538 München, Germany

ISSN 0172-6218

ISBN 3-540-44135-2 Springer-Verlag Berlin Heidelberg New York

Library of Congress Cataloging-in-Publication Data applied for.

Die Deutsche Bibliothek - CIP-Einheitsaufnahme

Chemistry of nanomolecular systems : toward the realization of molecular devices / T. Nakamura ... (ed.). -
Berlin ; Heidelberg ; New York ; Hong Kong ; London ; Milan ; Paris ; Tokyo : Springer, 2003
(Springer series in chemical physics ; 70)
(Physics and astronomy online library)
ISBN 3-540-44135-2

Springer-Verlag Berlin Heidelberg New York
a member of BertelsmannSpringer Science+Business Media GmbH

http://www.springer.de

Typesetting: Data conversion by LE-TEX Jelonek, Schmidt & Vöckler GbR, Leipzig
Cover concept: eStudio Calamar Steinen
Cover production: *design & production* GmbH, Heidelberg

Printed on acid-free paper SPIN: 10867331 57/3141/YL - 5 4 3 2 1 0

Preface

Recently, molecular electronics, especially that utilizing single molecules, has been attracting much attention. This is mainly because the theoretical limit is approaching in the present silicon-based technology, and the development of an alternative process is strongly desired. Single-molecule electronics is aimed at a breakthrough toward the next generation of computing systems. By designing and synthesizing highly functionalized molecules of nanometer size and incorporating these molecules into electrical circuits, we shall obtain much dense and high-speed processors.

The concept of single-molecule electronics was first introduced by Aviram and Ratnar in 1978. In the early 1980s, many groups all over the world had started research on molecular electronics. At that time, single-molecule manipulation techniques had not been born, and the research was mainly carried out on molecular films formed by the Langmuir–Blodgett technique, a wet process, and by molecular-beam epitaxy, a dry process. A number of prototypes of switching devices and logic gates were, however, reported in the 1980s. In the early 1990s, scanning probe microscopes became popular and researchers obtained a single-molecule manipulation and evaluation technique. It became possible to fabricate practical devices using single molecules or small numbers of molecules. Finally, at the end of the last century, an explosion in the research field of single-molecule electronics was witnessed.

In addition, studies of "biocomputing" started in the early 1980s and significant progress was achieved in the last century. The main purpose of those studies was to mimic biological functions and, more challengingly, to realize functions superior to biological functions artificially. For example, device prototypes based on bacteriorhodopsins and superparallel computing systems utilizing DNA have been extensively studied. Biocomputing systems, however, need not necessarily composed of biomaterials such as proteins and nucleic acids. The main goal of the research on biocomputing is to elucidate the mechanism of information processing in living organisms and to reproduce the computation processes in living systems artificially. Such systems can not be constructed from a single molecule, and a molecular-assembly approach based on supramolecular chemistry is highly desirable.

There are several scientific fields converging on the development of molecular electronics. Much research at present is aimed at developing single-molecule devices, while the molecular-assembly approach is quite important, especially, to those who are interested in non-Neumann-type computing architectures. The final goal of these research fields is still unclear. Presumably, structurally well-controlled nanosized architectures, such as molecule mimics of large-scale integrated devices, artificial biological systems, molecular machines, and advanced materials with enhanced nanoscale quantum physical properties, are candidates. But these goals are far away, and step-by-step basic research is inevitable. It is at least obvious that progress in the chemistry of molecules and molecular systems is one of the keys to realizing practical molecular-electronic devices and systems in the near future.

This book describes the efforts of young scientists in Japan to develop nanomolecular systems aimed toward molecular electronics. The contributions of these authors illustrate the frontiers of research along with a comprehensive introduction and a review of the state of the art in this field. The authors really know what is occurring at the frontiers and the book provides solid and valuable information on this. We hope the book will be useful not only to postgraduate students who have just entered the field of molecular electronics but also to busy researchers in related fields. The book is composed of three parts. The first part describes the synthesis of novel molecules for molecular nanosystems. The second part deals mainly with nanomolecular systems on solid surfaces and the evaluation of these systems with scanning probe microscopes. In the third part, the theory that forms a background to the experimental studies in molecular electronics is reviewed.

Part I: Synthetic Approaches to Nanomolecular Systems

To obtain structurally well-defined nanosized architectures, the approach from the viewpoint of synthetic chemistry is especially important. This approach involves the design and synthesis of the component molecules for molecular electronics. Part I of this book deals with the molecules and molecular systems that will be used in molecular electronics, with special emphasis on the relationship between the shapes of the molecules and nanoscale quantum phenomena.

In the first two chapters, switching molecules are reviewed. Such molecules respond to external fields or stimuli, such as heat, photons or a magnetic field, and subsequently change their structure. These molecules are expected to act as the smallest switches that will be applicable in logic gates. Chapter 1 focuses on the redox phenomena of organic π-electron systems associated with structural changes. The simultaneous response of molecules to both electric field and photons is also discussed. Chapter 2 deals with the photochemical switching of diarylethenes. Photochemical switching phenomena are usually detected by a change in color of the molecules. With the introduction of open-shell functional groups, such as stable neutral radicals, however, a change in the magnetic properties is induced by photons, which may be useful in constructing molecular-level information storage systems.

In Chap. 3, single-domain and single-molecule magnets are reviewed. Magnetic order is usually regarded as a three-dimensional bulk phenomenon. However, some metal clusters act as single-molecule magnets. Such molecules are considered to be one of the best candidates for the smallest information storage dots. In Chap. 4, the synthesis of the porphyrins and phthalocyanines frequently used in atomic-resolution STM studies is summarized. Atomic-resolution STM studies are essential for nanoscale fabrication. Porphyrins and phthalocyanines are extensively used in STM studies, because these molecules have specific molecular shapes and extended π-electron systems that allow them to be visualized easily. Recent work on atomic-resolution STM of these molecules is also reviewed.

Part II: Surface Molecular Systems

In order to utilize molecular-level phenomena in molecular devices, it is necessary to support the molecules on solid surfaces and connect them to macroscopic inputs and outputs. In general, a molecular device is engineered by "build-up" processes such as self-assembly and the synthesis of large molecules and supramolecules. On the other hand, macroscopic circuits and electrodes are fabricated by "top down" processes such as lithography that have been well developed in silicon technology. Recent advances in organic synthesis have allowed us to synthesize huge molecules whose sizes are several tens of nanometers. In addition, the interelectrode distance of planar nanojunctions has been reduced to 10 nm, which is small enough to connect single molecules. However, molecular devices based on the function of single molecules have not been successful to date, because our knowledge of molecule–surface interactions and of intermolecular interactions on surfaces is not sufficient to combine "built-up" and "top-down" processes. In part II, recent achievements in the investigation of surface molecular systems are presented

In the first two chapters, the formation of chemical bonds between molecules and solid surfaces is discussed. Investigation on the scale of single molecules has been achieved by the development of scanning probe microscopy. Chapter 5 focuses on the reactions of carboxylates with titanium dioxide surface. These are typical ionic reactions in which organic molecules form chemical bonds with an oxide surface. Oxides such as SiO_2 and Al_2O_3 are very important as insulating materials in electronic devices. Chapter 6 introduces self-assembled monolayers (SAMs) of organic molecules, created on a variety of surfaces. Some publications have reported that a diode and an FET have been made from SAMs, which acted as electronically active elements.

The last three chapters of part II give recent results about intriguing structures formed by organic molecules on solid surfaces. In some cases, organic molecules are able to move freely on the surface without molecule–surface chemical reactions. As a result, the molecules form striking structures through intermolecular interactions. Chapter 7 introduces the supramolecular chemistry of porphyrins and phthalocyanines adsorbed on solid surfaces. The highly ordered structures are controlled by competition among various interactions including surface adsorption, π-stacking, coordination, hydrogen bonds, and dipole interactions of chemical groups. In Chap. 8, nanoscale one-dimensional wires are described, including molecular nanowires of charge-transfer complexes fabricated by Langmuir–Blodgett methods. The electrical conductivity of these wires is also discussed. Chapter 9 introduces the control of dye aggregates in microscopic polymer matrixes. Dye molecules contained in polymer particles can be separated efficiently so as to prevent undesirable quenching due to intermolecular energy transfer. Such isolation is very important when one is interested in constructing molecular devices taking advantage of molecular excited states.

Part III: Theory for Nanomolecular Systems

This part provides the theoretical background for studies of the nanomolecular systems aimed toward molecular devices. Although the conduction mechanisms in bulk molecular solids and conducting polymers are well established, those in nanometer-scale objects are quite different. The experimental results reported so far for single molecules are still controversial. In Chap. 10, theoretical approaches to the electronic properties of nanomolecular systems and devices are reviewed. In particular, the conduction through metal clusters and single molecules between metal electrodes is discussed, which provides a basic understanding for designing single-molecule electronic devices. Several approaches are compared, including a method newly developed by the authors. In Chap. 11, quantum computation, which is one of the most promising computing methodologies for a post-Turing–Neumann-type computer, is reviewed. An overview of the theory and possible applications of quantum computation is given. The experimental approaches to quantum computation related to molecular-electronic devices are also described.

Sapporo,
August 2002

Ken-ichi Sugiura
Takuya Matsumoto
Hirokazu Tada
Takayoshi Nakamura

Contents

Part I Synthetic Approaches to Nanomolecular Systems

Part II Surface Molecular Systems

List of Contributors

Tomoyuki Akutagawa
Research Institute
for Electronic Science
Hokkaido University
N12W6, Kita-ku
Sapporo 060-0812, JAPAN
takuta@imd.es.hokudai.ac.jp

Kunio Awaga
Department of Chemistry
Graduate School of Science
Nagoya University
Furo, Chikusa
Nagoya 464-8602, JAPAN
awaga@mbox.chem.nagoya-u.ac.jp

Hiroki Higuchi
Division of Chemistry
Graduate School of Science
Hokkaido University
N10W8, Kita-ku
Sapporo 060-0810, JAPAN
higuchi@sci.hokudai.ac.jp

Masahiro Irie
Graduate School of Engineering
Kyushu University
6-10-1 Hakozaki, Higashi-ku
Fukuoka 812-8581, JAPAN
irie@cstf.kyushu-u.ac.jp

Takao Ishida
Institute of Mechanical Systems
Engineering (IMSE)
National Institute
of Advanced Industrial Science
and Technology (AIST)
1-2-1 Namiki
Tsukuba, Ibaraki 305-8564, JAPAN
t-ishida@aist.go.jp

Olaf Karthaus
Chitose Institute
of Science and Technology
Bibi
Chitose 066-8655, JAPAN
karthaus@photon.chitose.ac.jp

Tomoji Kawai
The Institute of Scientific
and Industrial Research,
Osaka University
8-1 Mihogaoka, Ibaraki
Osaka 567-0047, JAPAN
kawai@sanken.osaka-u.ac.jp

Kenji Matsuda
Graduate School of Engineering
Kyushu University
6-10-1 Hakozaki, Higashi-ku
Fukuoka 812-8581, JAPAN
kmatsuda@cstf.kyushu-u.ac.jp

Takuya Matsumoto
The Institute of Scientific
and Industrial Research
Osaka University
8-1 Mihogaoka, Ibaraki
Osaka 567-0047, JAPAN
matsumoto@sanken.osaka-u.ac.jp

Tsutomu Miyashi
Department of Chemistry
Graduate School of Science
Tohoku University
Aoba-ku
Sendai 980-8578, JAPAN
miyashi@org.chem.tohoku.ac.jp

Takayoshi Nakamura
Research Institute
for Electronic Science
Hokkaido University
N12W6, Kita-ku
Sapporo 060-0812, JAPAN
tnaka@imd.es.hokudai.ac.jp

Jun-ichi Nishida
Department of Electronic Chemistry
Interdisciplinary Graduate School
of Science and Engineering
Tokyo Institute of Technology
Midori-ku
Yokohama 226-8502, JAPAN
jnishida@echem.titech.ac.jp

Hiroshi Onishi
Kanagawa Academy of Science
and Technology
KSP E-404, Sakado, Takatsu
Kawasaki 213-0012, JAPAN
oni@net.ksp.or.jp

Kenichi Sugiura
Department of Chemistry
Graduate School of Science
Tokyo Metropolitan University
1-1 Minami-Ohsawa, Hachi-Oji
Tokyo 192-0397, JAPAN
sugiura@porphyrin.jp

Takanori Suzuki
Division of Chemistry
Graduate School of Science
Hokkaido University
N10W8, Kita-ku
Sapporo 060-0810, JAPAN
tak@sci.hokudai.ac.jp

Keiji Takeda
Division of Chemistry
Graduate School of Science
Hokkaido University
N10W8, Kita-ku
Sapporo 060-0810, JAPAN
takeda-k@sci.hokudai.ac.jp

Shigeki Takeuchi
Research Institute
for Electronic Science
Hokkaido University
N12W6, Kita-ku
Sapporo 060-0812, JAPAN
takeuchi@es.hokudai.ac.jp

Takashi Tsuji
Division of Chemistry
Graduate School of Science
Hokkaido University
N10W8, Kita-ku
Sapporo 060-0810, JAPAN
tsuji@sci.hokudai.ac.jp

Satoshi Watanabe
Dept. of Materials Engineering
Graduate School of Engineering
The University of Tokyo
7-3-1 Hongo, Bunkyo-ku
Tokyo 113-8656, JAPAN
watanabe@cello.t.u-tokyo.ac.jp

Yoshiro Yamashita
Department of Electronic Chemistry
Interdisciplinary Graduate School
of Science and Engineering
Tokyo Institute of Technology
Midori-ku
Yokohama 226-8502, JAPAN
yoshiro@echem.titech.ac.jp

Takashi Yokoyama
National Institute
for Materials Science
2268-1 Shimo-shidami, Moriyama-ku
Nagoya 463-0003, JAPAN
YOKOYAMA.Takashi@nims.go.jp

Part I

Synthetic Approaches to Nanomolecular Systems

1 Dynamic Redox Systems: Toward the Realization of Unimolecular Memory

Takanori Suzuki, Hiroki Higuchi, Takashi Tsuji, Jun-ichi Nishida, Yoshiro Yamashita, Tsutomu Miyashi

Summary. "Dynamic redox systems" is the name given to a certain class of compounds that can be reversibly converted into the corresponding charged species upon electron transfer (ET) but accompanied by drastic structural changes and/or covalent bond making/breaking. Although transformation between the neutral and charged species proceeds quantitatively in most cases, their electrochemical processes are irreversible owing to the chemical reactions followed by ET. This situation endows the "dynamic redox pairs" with very high bistability, which is a prerequisite in order to construct the molecular response systems.

This chapter first describes the principle of the "dynamic redox properties", and then classifies the compounds into several categories from the viewpoint of the type of structrual changes. After the properties of representative molecules in each category have been described in detail, their expected use as unimolecular devices is commented on in the final part of the chapter.

1.1 Introduction

Organic redox systems have been attracting considerable attention owing to their intriguing properties, such as electrical conduction and ferromagnetism, in the field of materials chemistry [1]. Tetrathiafulvalene (TTF) [2] and tetracyanoquinodimethane (TCNQ) [3] (Fig. 1.1) are the representative electron donor and acceptor, respectively, which have been studied over a number of years. The first molecular metal [4], which initiated a vast study of organic conductors, was obtained by the combination of these two substances. Hundreds of new compounds have been designed and prepared as new multistage redox systems to realize higher conductivity or even superconductivity; most of these compounds exhibit a very slight structural change upon single electron transfer (ET). There are, however, a certain number of molecules that exhibit reversible, dynamic changes in their geometrical and electronic structures (hereafter, these molecules will be called "dynamic redox systems"). The latter compounds were judged to be inappropriate for producing organic conductors owing to the thermodynamic instability of their ion radicals [5].

However, when the library of these organic redox couples is reviewed on the basis of the criteria for molecular switches or response systems, dynamic

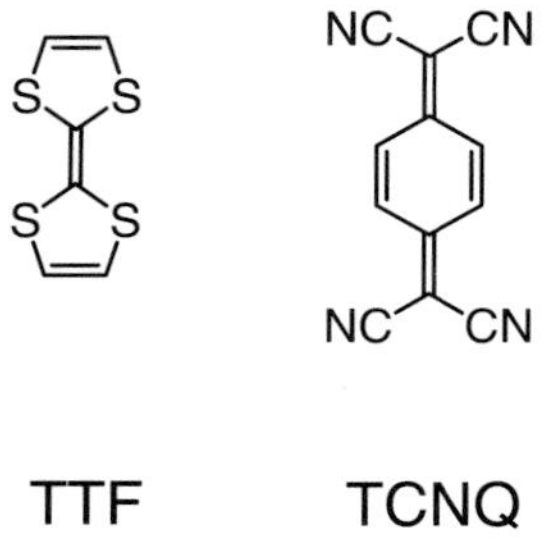

TTF TCNQ

Fig. 1.1. Structural formula of representative electron donor and acceptor

redox systems are seen to be promising candidates for realizing molecular devices. Electrochromism is one of the easily accessible applications of these molecules [6,7], and the high electrochemical bistability observed in some cases is quite suitable for making use of these compounds as molecular switches or memories. Considering the rapid development of surface-modifying methods [8] and of STM technology, it could become possible to induce a redox reaction of a single molecule on a surface by using an STM tip. Such a technique could open up a new way toward super-density memories (10^{12} devices/mm^2) by using each molecule of a set of molecules as a digit. The validity of this idea is supported by a very recent success in detecting a conformational change of a single molecule on a surface by topographic STM images [9].

In view of the important contribution expected in the field of molecular electronics, the detailed features of dynamic redox systems will be documented in this chapter. Molecular switches triggered by light are discussed in Chap. 2. In order to concentrate on the less well-developed small organic molecules of this type and to avoid overlap with existing well-written review articles [10–12], supramolecular systems such as catenanes and rotaxanes, and metal complexes are excluded. This chapter focuses on the basic concepts of dynamic redox behavior by referring early examples and attempts to predict where these studies will go in the future.

1.2 Redox Systems that Exhibit Drastic Geometrical Changes upon ET

The dynamic redox systems can be divided into two groups: one consists of molecules that show drastic structural changes upon ET, and the other contains compounds that exhibit reversible covalent-bond making/breaking accompanied by geometrical changes. In this section, the redox systems belonging to the first category are discussed. These members are further divided into four subgroups depending on the structures of π-frameworks adopted by the oxidized and reduced species.

1.2.1 Folded–Planar Structural Change

This type of geometrical change was first demonstrated by X-ray structural analyses conducted on both the neutral and the cation-radical forms of tetraalkylhydrazine **1** (Fig. 1.2) [13]. The pyramidal N atoms of the neutral species become flattened upon 1e-oxidation, and this is accompanied by considerable shortening of the N–N distance due to the increased bonding character in the cation radical. Such a geometrical change requires large vibrational reorganization energy for the tetraalkylhydrazine derivatives [14]. Thus, the electron exchange between the neutral and cation-radical species is very slow, even when it is an intramolecular process as in **2** [15,16].

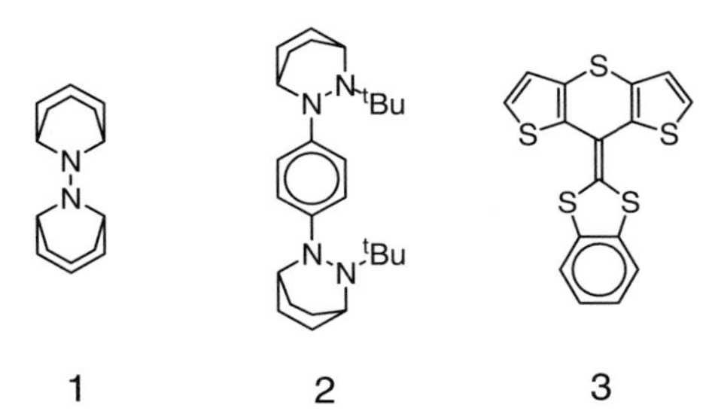

Fig. 1.2. Structural formula of electron donors

In the case of the thiopyranylidene-1,3-dithiol-type electron donor **3** showing intramolecular S ·· S contacts, its molecular structure is also changed upon 1e-oxidation [17]. The butterfly-shaped geometry in neutral **3** is a result of the repulsive interaction between S atoms (3.05 – 3.09 Å). However, removal of an electron makes the molecular geometry planar. The S ·· S distances are much shorter in the cation radical (2.86 – 2.87 Å). This value is intermediate between that of S–S single bonds (average 2.07 Å) and the sum of the van der Waals radii (3.70 Å), indicating the bonding character in these contacts. This may be the reason for the very small twisting around the central olefinic bond in $\mathbf{3}^{+\bullet}$ (4.6°).

1.2.2 Planar–Twisted Structural Change

In connection with the planar olefinic bond in $\mathbf{3}^{+\bullet}$, it is interesting to start this subsection with the controversy over the theoretical predictions for the geometries of olefin dications ($\mathbf{4}^{2+}$, $^{+}CX_2$–CX_2^{+} (Fig. 1.3)) [18,19]. According to an ab initio calculation, the parent ethylene dication ($\mathbf{4a}^{2+}$: X = H) adopts the perpendicular geometry, which is more stable than the planar geometry by 30 kcal mol^{-1}. In contrast, a planar structure is favored in the tetrafluoro ($\mathbf{4b}^{2+}$: X = F) and tetrahydroxy ($\mathbf{4c}^{2+}$: X = OH) derivatives, and the twisted forms were predicted to be 8 – 12 kcal mol^{-1} higher in energy than the planar dications. Such a discrepancy seems to be a result of the subtle balancing of π-conjugation effects and hyperconjugation of the C–X σ bond. Thus, the perpendicularly twisted dication is again more stable in the case of $\mathbf{4d}^{2+}$ (X = SH) because the hyperconjugation effects of C–S bonds are more effective than those of C–O or C–F bonds. The geometrical features were experimentally studied on the stable tetraaminoethylene dication ($\mathbf{4e}^{2+}$: X = NMe_2) by X-ray analyses [20]. Each of the two molecular halves of the dication is planar, and the short C–N distances (average 1.31 Å) are indicative of charge delocalization. The twisting angles (θ) are 76° and 67° for the Cl^- and Br^- salts, respectively, and are close to the value calculated by the MNDO method (62°).

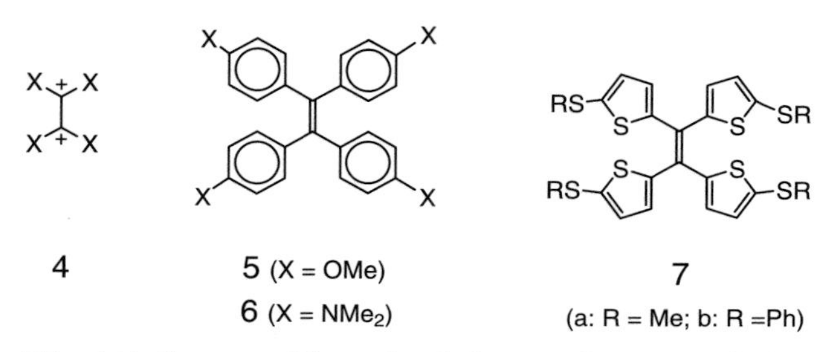

Fig. 1.3. Structural formula of electron-donating ethylenes

X-ray studies were also carried out on dications having four aryl groups with electron-donating substituents. The twisting angle of the tetrakis(4-methoxyphenyl)ethylene dication $\mathbf{5}^{2+}$ is 41° for the I_2Cl^- salt [21] and 62° for the mixed counter-anion salt containing $SbCl_6^-$ and Cl^- [22]. According to voltammetric analyses, the first (E_1^{ox}) and second (E_2^{ox}) oxidation potentials of **5** are very close to each other, probably owing to the dynamic structural and electronic changes upon electron extrusion. This idea is supported by the twisted geometry ($\theta = 30.5°$ by X-ray) of the cation-radical species of **5**, isolated as an $SbCl_6^-$ salt. In the case of the dimethylamino derivative **6**, with stronger donating properties, E_2^{ox} becomes even lower than E_1^{ox}, thus showing a single pair of sharp peaks in the voltammogram for the $\mathbf{6}/\mathbf{6}^{2+}$ process [23]. Although $\mathbf{6}^{+\bullet}$ could be detected by electron spin resonace (ESR), its steady-state concentration is low owing to the easy disproportionation. Thus, the spectroelectrogram obtained upon electrochemical oxidation of **6** (λ_{max} 295, 355 nm) to $\mathbf{6}^{2+}$ (490, 540, 690 – 750 nm) showed a clear isosbestic point. A PM3 calculation predicted a preference for the twisted conformation in the oxidized species ($\theta = 33.8°$ for $\mathbf{6}^{+\bullet}$ and 80.6° for $\mathbf{6}^{2+}$).

A completely bisected geometry ($\theta = 90.0°$) was demonstrated by the 5-methylthio-2-thienyl-substituted dication salt $\mathbf{7a}^{2+}(ClO_4^-)_2$ [24], in which $\mathbf{7a}^{2+}$ is located on the crystallographic $\bar{4}$ axis in the space group $P4/n$ ($Z = 2$). This geometry is intrinsic to this species but is not due to the crystal packing force, because X-ray analyses of salts with different counteranions or different substituents revealed a similar perpendicular structure: results have been obtained for the $\mathbf{7a}^{2+}(\mathrm{SbCl}_6^-)_2$ salt ($\theta = 89.4°$; space group $Fddd$, $Z = 8$) [25] or $\mathbf{7b}^{2+}(\mathrm{SbCl}_6^-)_2$ salt ($\theta = 86.2°$; space group $C2/c$, $Z = 4$) [26]. Although the above-mentioned examples indicate that the twisted geometry is rather common in isolable dications, not only electronic but also steric factors need to be considered to account for such a preference. Many of the olefins affording stable dications belong to the violene/cyanine hybrid systems [7,23,27], where each of the two positive charges is delocalized over half of the molecule. To maximize the conjugative effects, the shape of the half units in the dication becomes more extended than in the neutral state, thus increasing the steric bulkiness, which facilitates the twisting around the central C–C bond.

It is still unclear if the cation-radical intermediates adopt a twisted structure similar to that of the dications. However, it is reasonable to assume that rotation around the central bond facilitates the extrusion of a second electron from the cation radical, since diarylmethyl radicals usually exhibit stronger donating properties than do closed-shell ethylenes. So, in systems involving a twisted cation radical, the steady-state concentration of this intermediate is negligible, which is one of the conditions that is favorable for constructing the electrochromic systems because of the lack of destructive side reactions from the reactive cation radicals.

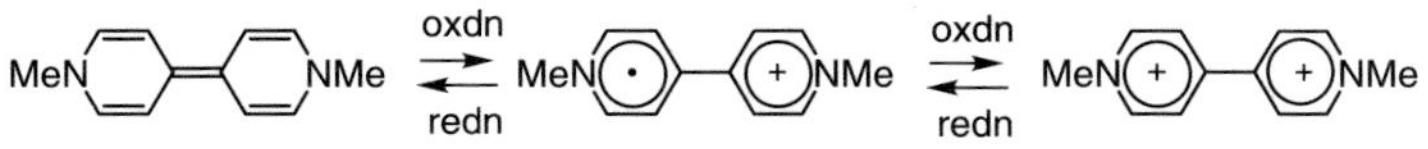

MeN NMe ⇌ (oxdn / redn) MeN NMe ⇌ (oxdn / redn) MeN NMe

8 $8^{+\bullet}$ 8^{2+}

Fig. 1.4. Redox reaction of methylviologen

It should be noted here that there is another class of dications whose structures are completely planar. The methylviologen dication $\mathbf{8}^{2+}$ (Fig. 1.4) and its derivatives are well-known organic dyes that undergo reversible two-stage one-electron reduction, and all three stages ($\mathbf{8}^{2+}$, $\mathbf{8}^{+\bullet}$, $\mathbf{8}$) were proven to be planar by X-ray analyses [12,28]. It is likely that other sterically undemanding dications can also adopt a planar structure. In this connection, the screwed geometry of the TTF dication ($\theta = \mathrm{ca.}\ 60°$) [29,30] may be accounted for by electronic effects such as hyperconjugation, because $\mathrm{TTF}^{+\bullet}$ is planar as has been observed in many conducting organic solids.

1.2.3 Twisted–Twisted Structural Change

1,1,4,4-tetrakis(4-dimethylaminophenyl)-1,3-butadiene **9** (Fig. 1.5) is a vinylogue of the olefin **6**. This compound belongs to the category of open-chain violenes [31], which are representative electrochromic materials. According to the PM3 calculation, the diene unit remains planar during the interconversion between **9** and $\mathbf{9}^{2+}$ [23]. This may be also true for the TTF vinylogues **10** [32,33]. However, introduction of two aryl substituents at the 2 and 3 positions induces dynamic structural changes upon ET [34]. According to X-ray analyses, **11** adopts a twisted geometry with a torsion angle of 61.2° for the diene unit, whereas 1e-oxidation induces complete planarization (torsion angle 180°) of the four-carbon unit, to which the two aryl groups are attached with a dihedral angle of 83.9°. Such a drastic motion by rotation around the C–C bonds is more prominent when the diene unit is framed into a ring structure, as in 1,2-bis(exomethylene) compounds such as **12** [35].

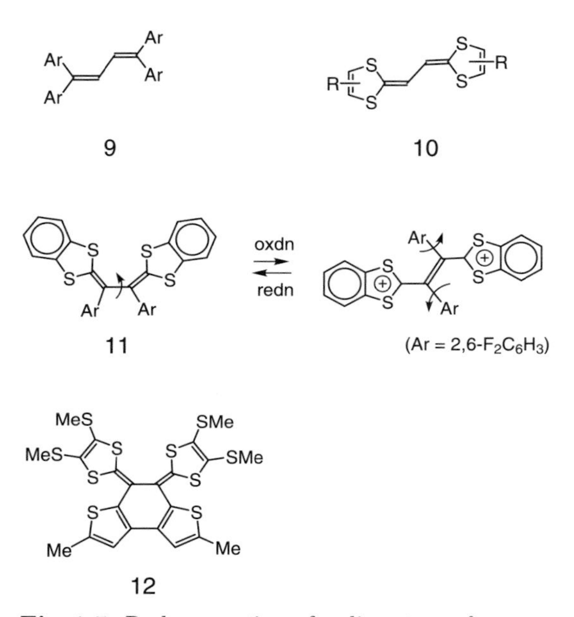

Fig. 1.5. Redox reaction of a diene-type donor

Detailed studies of the dynamic properties were conducted on 5,6-dimethylene-1,2,4-trithiane **13** (Fig. 1.6) [36], which has the same chromophore as **9**. This red crystalline material (λ_{max} 394sh, 310, 276sh nm) is reversibly interconverted with the deep purple dication $\mathbf{13}^{2+}$ (677, 593sh, 561, 434, 303, 265 nm) upon 2e-oxidation. The presence of an isosbestic point in the spectroelectrogram for the $\mathbf{13}/\mathbf{13}^{2+}$ process shows a clean transformation under electrochemical conditions, as well as a negligible steady-state concentration of $\mathbf{13}^{+\bullet}$. X-ray analyses indicate that geometrical changes between differently twisted forms occur upon ET. The torsion angle of the diene unit in neutral **13** is 51.0°, whereas the corresponding angle in $\mathbf{13}^{2+}$ is 0°. Two diarylcarbenium units in the dication rotate against the 6-membered ring with dihedral angles of 62.2° and 62.7°. Similar structural changes are expected in the 5-membered-ring analogue 4,5-dimethylene-1,3-dithiole **14** (λ_{max} 385sh, 315, 290sh nm), which can be interconverted with the deeply colored dication **14**$^{2+}$ (703, 600, 544, 440, 358, 293 nm), and thus consists of an analogous chromic system [37]. Hexaaryl-1,3-butadiene **15** is another example that exhibits a twisted–twisted motion upon ET [38]. It is noteworthy that the dication **15**$^{2+}$ adopts a *cis*-configuration in the crystalline phase, suggesting an attractive interaction between two diarylcarbenium units because of the face-to-face overlap.

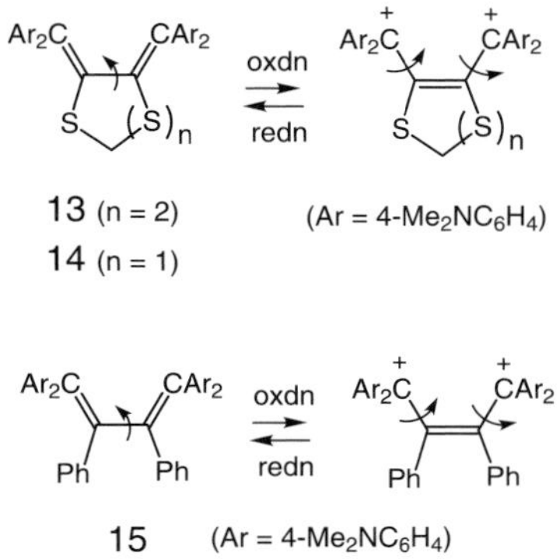

Fig. 1.6. Redox reactions of diene-type donors

1.2.4 Folded–Twisted Structural Change

p-Quinobis(1,3-dithiole) **16** (Fig. 1.7) is a π-extended derivative of TTF showing very strong donating properties [39], which seemingly adopts a planar geometry, as in the case of **17** which is fused with electron-withdrawing 5-membered heterocycles [40]. When a benzene or naphthalene ring is substituted for one of the two heterocycles in **17**, the steric repulsion between S atoms and peri-hydrogens may induce a butterfly-shaped deformation of the molecule, as shown by X-ray analysis of **18a** [41]. It is interesting to note that 1e-oxidation of **18** does not induce any subtle structural change, and that cation-radical salts (e.g. **18b**$^{+\bullet}$) with folded structure show high conductivity or even metallic behavior [42,43]. Much severer steric repulsion is expected when two benzene rings are fused to both sides of **16**. Thus, the dibenzo analogue **19** (Fig. 1.8) was shown to have a folded structure by X-ray analysis, whereas its dication salt **19**$^{2+}$ has a planar anthracene unit, to which the two 1,3-dithiolium rings are attached with a dihedral angle of 86.0° (folded–twisted motion) [44].

Similar structural changes were observed in dibenzo analogues of TCNQ **20** (Fig.1.9) [45] that undergo one-wave 2e-reduction. X-ray analyses indicate a butterfly-shaped geometry for the neutral acceptor (R = H) [46] and for

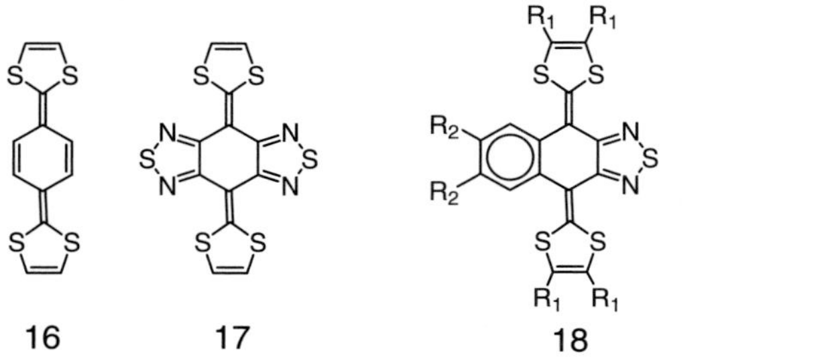

16 17 18

(a: R_1,R_1 = benzo, R_2 = H;
b: R_1,R_1 = SCH_2CH_2S; R_2,R_2= benzo)

Fig. 1.7. Structural formula of quinoid-type donors

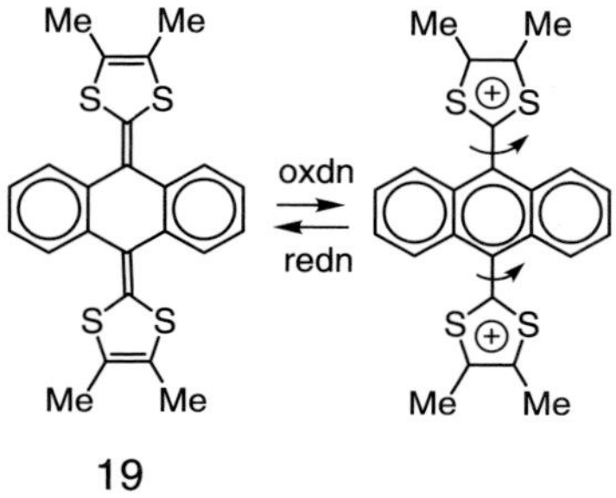

19

Fig. 1.8. Redox reactions of a quinoid-type donor

its CT complexes (R = Cl) [47]. However, detailed electrochemical analyses suggest that the dianion has a planar anthracene unit, to which two dicyanomethylides are attached nearly perpendicularly [45,48]. A similar phenomenon was observed in the pentacenoquinodimethane derivative [49] and in a more sophisticated recent example containing two anthraquinodimethane units [50]. Such structural changes are again induced by the steric repulsion between the cyano groups and peri-hydrogens in both the neutral and the reduced species. Thus, the heterocyclic analogues (**21**–**23**) with no peri-hydrogens are reduced to the dianions in a stepwise manner via the isolable anion radicals [51–53] because they have planar geometries in the neutral state [53–55], without subtle geometrical changes upon ET.

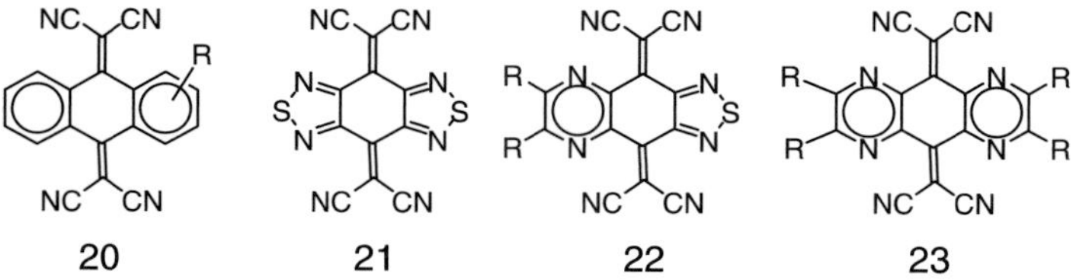

20 21 22 (R = H, Me, Ph) 23 (R = Me, Ph)

Fig. 1.9. Structural formula of quinoid-type acceptors

According to X-ray analysis, bianthraquinodimethane **24** (Fig. 1.10) adopts a doubly folded structure [56]. Electrochemical analysis showed a one-wave 2e-reduction process for **24**, thus suggesting a twisted conformation for its dianion. Bianthrone, **25**, the quinone analogue of **24**, is known as a representative thermochromic compound that exhibits a drastic color change from yellow to deep green [57]. The origin of the chromism was proven to be isomerization between the folded conformer **25F** [58] and the twisted one **25T** [59,60]. Thus, **25** can adopt both folded and twisted geometries in the neutral state, and external stimuli such as heat cause an increase in the amount of the metastable conformer **25T** so as to induce the chromic phenomenon. In contrast, the twisted form is the more stable conformation for the reduced species; thus ET also induces a folded–twisted structural change in **25** [61].

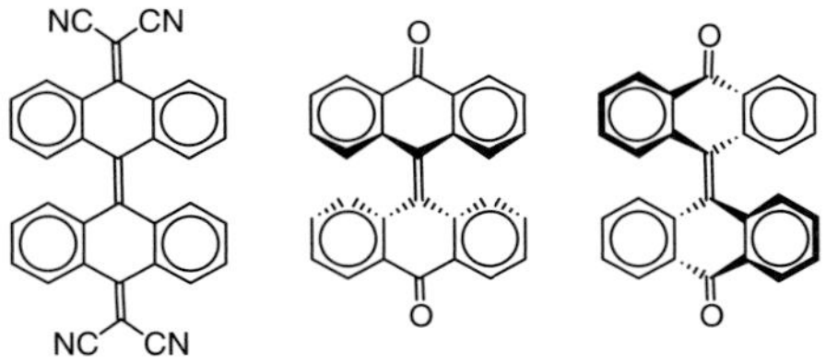

Fig. 1.10. Structural formula of electron accepting overcrowded ethylenes

1.3 Redox Systems that Undergo Reversible Bond Making/Breaking upon ET

The compounds described in this section have a higher electrochemical bistability than do the molecules discussed in Sect. 1.2 because the electronic structures of the neutral and ionic species are altered dramatically by forming/cleaving a covalent bond. It has been known for years that nonconjugated chromophores in organic molecules interact electronically so as to show UV–visible absorption in a longer-wavelength region than in the unichromophoric reference compounds [62]. Such a behavior can be accounted for by through-bond and/or through-space coupling of molecular orbitals [63]. In some cases, such interactions favor the formation of a new bond (a 2-center 3-electron bond) [64] upon redox reactions. Diamines and dithioethers have been used to study the interaction between two lone pairs, which will be briefly commented on in Sect. 1.3.1. Organic π-electron systems are described in the next two subsections by grouping them into two categories (C–C bond making upon ionization and C–C bond breaking upon ionization).

1.3.1 Reversible Bond Making/Breaking Between Heteroatoms upon ET

An X-ray structural analysis of diazabicyclo[3.3.3]undecane fused with a naphthalene unit, **26** (Fig. 1.11), has revealed that the two nitrogen atoms adopt an unusual planar geometry, which might account for its very low HOMO level (7.3 eV) as well as its low-energy electronic transition (λ_{max} 380 nm) [65]. Upon extrusion of one or two electrons from **26**, a new σ bond is formed between the nitrogens [66]. The naphthalene π-system of this molecule does not play an important role, since diazabicyclo[4.4.4]tetradecane **27**, without any aromatic nucleus, also exhibits a similar N–N bonding upon oxidation [67].

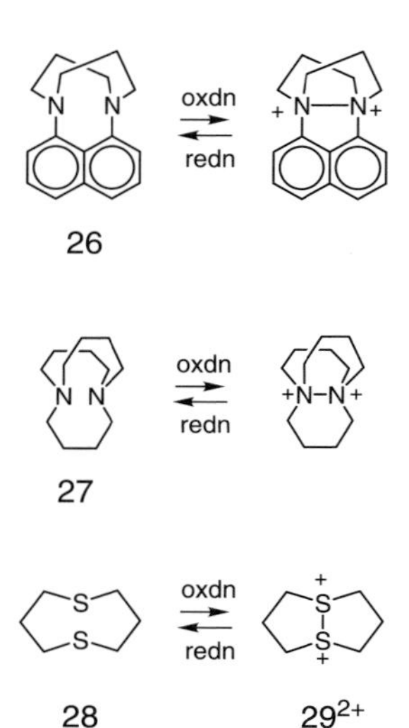

Fig. 1.11. Structural formula of electron donors exhibiting interheteroatom bonding

Cyclic dithioethers such as **28** are another class of compounds that are attracting considerable attention from the viewpoint of transannular interaction and bonding [68]. X-ray analysis of the $\mathbf{29}^{2+}$ salt showed the presence of an S–S σ bond unambiguously [69], and the chemistry of this compound, including higher-chalcogen analogues, has been thoroughly investigated, as summarized in recent review articles [70,71].

1.3.2 π-Electron Systems that Undergo C–C Bond Making upon Ionization

1,3-Dimethylenecyclobutane is an attractive framework that would be transformed into a highly strained bicyclo[1.1.0]butane skeleton by making a C–C bond. This behavior was successfully demonstrated by interconversion between the Weitz-type electron donors **30** and the dications $\mathbf{31}^{2+}$ (Fig. 1.12) [72–74], and further exploited by examining fulvene-type hydrocarbons [75] and bis(quinone methide)-type acceptors [76]. Although the complete redox cycle is reversible, both of the ET reactions of **30** and **31**$^{2+}$ are irreversible owing to the difference in their structures. For example, the oxidation of **30a** occurs at $+0.17$ V vs. Ag/Ag^+, while the reduction of **31a**$^{2+}$ occurs at -0.38 V, thus giving a large peak separation of 0.55 V in the cyclic voltammogram. A one-wave 2e-oxidation peak is another characteristic of this redox pair. These two features in the voltammogram are diagnostic of reversible C–C bond making/breaking upon ET.

Transannular C–C bonding was also examined by using dibenzo-TTF analogues with a cycloalkane ring inserted [77]. The stronger donating properties (E^{ox} $+0.55$ V vs. SCE) of the cyclooctane derivative **32a** (Fig. 1.13) than of the fully conjugated dibenzo-TTF (E^{ox} $+0.74$ V) are indicative of significant interaction between the two chromophores separated by the alicyclic ring. Upon 2e-oxidation of **32a**, the dication **33a**$^{2+}$ with a bicyclo[3.3.0] skeleton was isolated as a stable salt, whose structure was confirmed by X-ray analysis. Upon 2e-reduction of **33a**$^{2+}$, the diolefin **32a** was regenerated quantitatively. Similarly, the cyclodecane derivative **32b** undergoes transannular C–C bonding to give the dication **33b**$^{2+}$. However, **33c**$^{2+}$, with a bicyclo[2.2.0]hexane framework, was not produced, probably owing to its larger strain energy. In this connection, it is interesting to note that the dihydrophenalene derivative **34** undergoes oxidative C–C bond making to form the dication **35**$^{2+}$ with a cyclopropane ring [78]. The absence of the delocalized dication **34**$^{2+}$ can be accounted for by the nonplanarity of the π-framework due to steric repul-

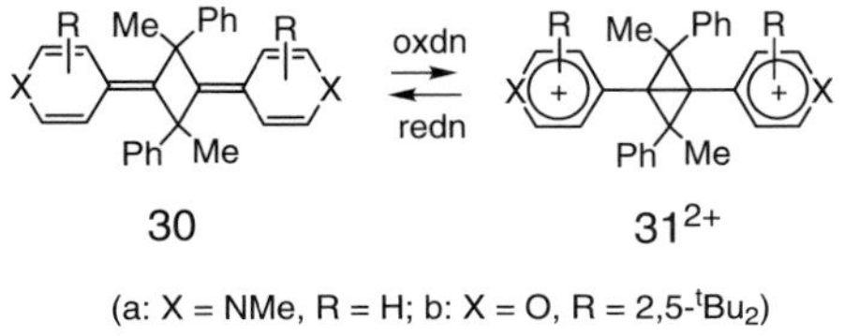

(a: X = NMe, R = H; b: X = O, R = 2,5-^{t}Bu$_2$)

Fig. 1.12. Redox reactions of dimethylenecyclobutanes

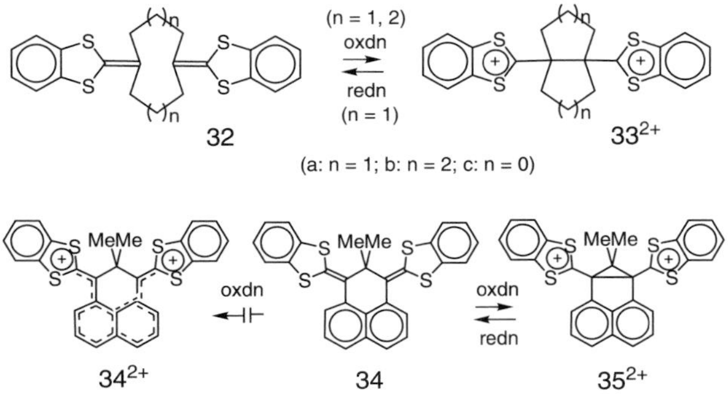

(a: n = 1; b: n = 2; c: n = 0)

Fig. 1.13. Dynamic redox properties of TTF analogues

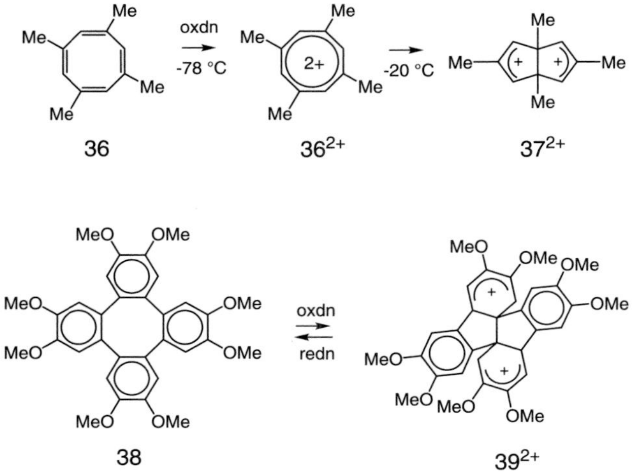

Fig. 1.14. Redox reactions of cyclooctatetraenes

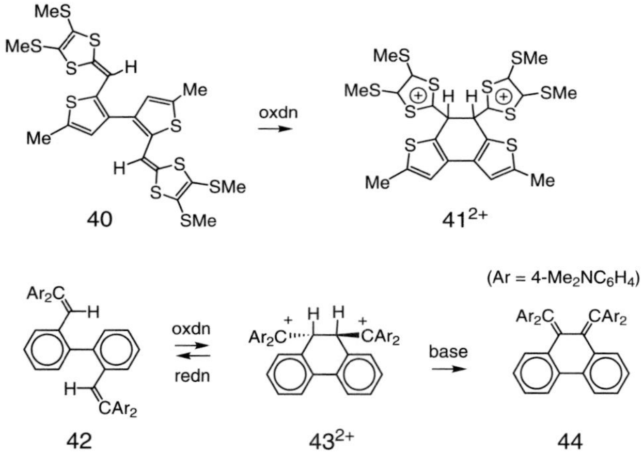

Fig. 1.15. Redox reactions of diene-type donors

sion, as well as non-Kekulé-type conjugation of 1,8-naphthoquinodimethane. A preference for a transannularly bonded structure rather than a delocalized structure was previously observed in the COT dication **36**$^{2+}$ (Fig. 1.14) [79]. Although NMR analysis at −78 °C indicated aromatic character for **36**$^{2+}$, this molecule isomerized irreversibly to **37**$^{2+}$ with a bicyclo[3.3.0]octadiene skeleton. This type of C–C bonding in the dication also occurs in the tetrabenzo analogue of COT, **38**, to form the bicyclic dication **39**$^{2+}$, which regenerates **38** upon 2e-reduction [80].

Another type of dynamic redox systems was designed by substituting electroactive groups at the 2,2′-positions of biaryls. 3,3′-Bithienyl **40** (Fig. 1.15), containing two electron-donating 1,3-dithiol-2-ylidene groups, was effectively transformed into the bridged dication $\mathbf{41}^{2+}$ by C–C bond making upon oxidation. This dication is unstable under the reaction conditions used and is easily deprotonated to the conjugated diene-type donor **12** [35]. In contrast, another bridged dication, $\mathbf{43}^{2+}$, could be isolated as a stable salt by 2e-oxidation of the biphenyl-type donor **42** with two bis(4-dimethylaminophenyl)ethenyl units attached [81]. The surprising stability of the 1,4-dication, $\mathbf{43}^{2+}$, can be accounted for by the steric shielding of the acidic C–H protons, as well as an unfavorable arrangement of the C–H σ bond and the vacant p-lobes for producing the deprotonated diene **44**.

All the above-mentioned examples are of electron donors that are transformed into bridged dications by C–C bonding upon 2e-oxidation (Fig. 1.16). A similar redox cycle can be considered for pairs consisting of an electron acceptor and the corresponding bridged dianion [76]. 2,2-Bis(dicyanovinyl)biphenyl is a colorless material but substitution with two electron-donating dimethylaminophenyl groups induces strong absorption in the visible region (λ_{max} 455 nm) by push–pull type conjugation. The dye **45** (Fig. 1.17) underwent reversible C–C bond making to form the faintly colored dianion $\mathbf{46}^{2-}$, thus constructing another class of electrochromic system based on dynamic redox behavior. Although the dianion could not be isolated as a salt, regioselective formation of the *trans*-isomer was confirmed by X-ray analysis of the diprotonated species [82]. In the case of bis(quinomethide) **47**, with

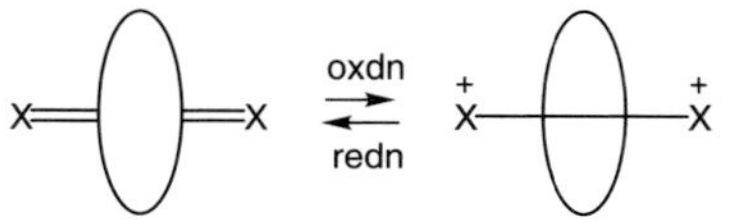

Fig. 1.16. Schematic formula of dynamic redox properties

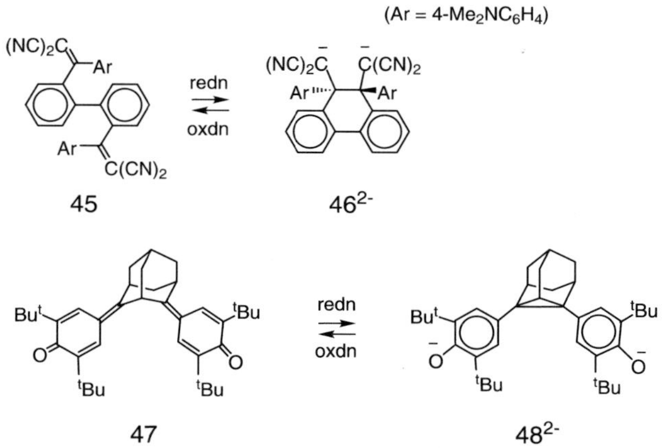

Fig. 1.17. Dynamic redox properties of diene-type acceptors

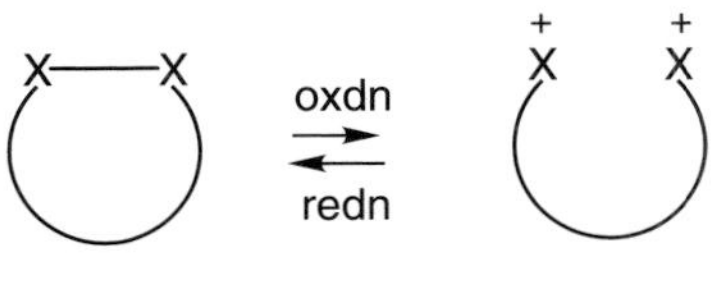

Fig. 1.18. Schematic formula of dynamic redox properties

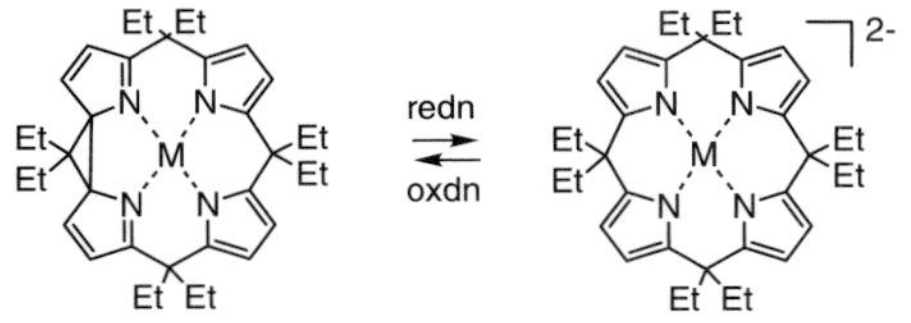

Fig. 1.19. Redox reactions of acceptors

a rigid adamantane framework, C–C bonding upon 2e-reduction resulted in formation of the highly strained dehydroadamantane skeleton, as deduced by NMR spectroscopy of the dianion $\mathbf{48}^{2-}$ and its TMS ether [83].

Compared with their cationic counterparts, the dynamic redox systems involving anionic species have been less exploited, probably owing to the difficulty of isolating the anions as salts. So, detailed studies of C–C bond breaking upon ionization have been conducted for redox pairs involving an isolable dication (Fig. 1.18), as shown in the next subsection, although there are also several reports of intriguing redox pairs containing polyanions such as cyclic tetrapyrroles (**49** and $\mathbf{50}^{2-}$, Fig. 1.19) [84,85] and bis(fluorenide) (**51** and $\mathbf{52}^{2-}$) [86] and its analogues containing a C_{60} framework [87,88].

1.3.3 π-Electron Systems that Undergo C–C Bond Breaking upon Ionization

Mesolytic C–C bond fission proceeds more easily than homolysis. The average activation energy for the fragmentation of 1,2-diarylethane cation radicals was reported to be 23 kcal mol^{-1} lower than that for homolysis [89]. Multiple introduction of aryl substituents around the C–C bond induces bond elongation [90,91] by steric interaction [92,93], thus making its fission much easier. Hexaphenylethane **53** (Fig. 1.20), the ultimate target of this design concept, had been a compound with a lot of riddles [94], and could be isolated only when bulky substituents were present on the benzene nuclei [95,96] to prevent the formation of the *α,p*-dimer **54**. Another way to stabilize the

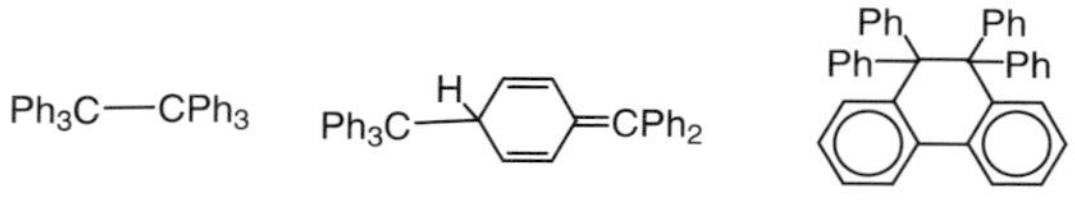

Fig. 1.20. Structural formula of hexaphenylethane derivatives

hexaphenylethane skeleton is by clamping two benzene rings. Thus, 9,9,10,10-tetraphenyl-9,10-dihydrophenanthrene **55** has been known as a stable crystalline material [97], and an important prototype of dynamic redox systems was designed on the basis of this framework.

By introduction of electron-donating functional groups such as Me_2N(**a**) or MeO (**b**) on the phenyl rings, the central C–C bond in **56** (Fig. 1.21) was shown to cleave very readily upon 2e-oxidation. The oxidized species are the dications **57**$^{2+}$ with a biphenyl skeleton, which were isolated as stable salts. Owing to the formation of triarylmethane-type chromophores upon oxidation, this redox pair exhibits electrochromism with a vivid change in color (λ_{max} 268 nm for **56a**; 661 and 604 nm for **57a**$^{2+}$) [98,99]. Owing to the C–C bond making/breaking as well as dynamic geometrical changes upon ET, E^{ox} of **56** and E^{red} of **57**$^{2+}$ differ by more than 1 V, although the high-yield interconversion indicates that they can be considered as a sort of "reversible" redox pair.

When the biphenyl skeleton of **57**$^{2+}$ was replaced by a diphenyl ether moiety, 2e-reduction no longer gave the hexaphenylethane derivative **58** (Fig. 1.22) with a 7-membered ring. Instead, the 9-membered cyclic peroxides **59** were isolated in good yield upon reduction of bis(xanthenylium)-type dications **60**$^{2+}$ under air. It is interesting to note that **59** regenerated **60**$^{2+}$ effectively through an "oxidative deoxygenation" reaction, a rather paradoxical mode of redox reaction. The pair of **59** and **60**$^{2+}$ represents a new type of chromic system that reversibly traps and expels O_2 gas on electrochemical input [100]. When a binaphthyl skeleton is used as the arylene spacer that connects the two chromophores in the dication, the chiroptical properties can also be modified electrochemically. Thus, the dihydro[5]helicenes **61** and the

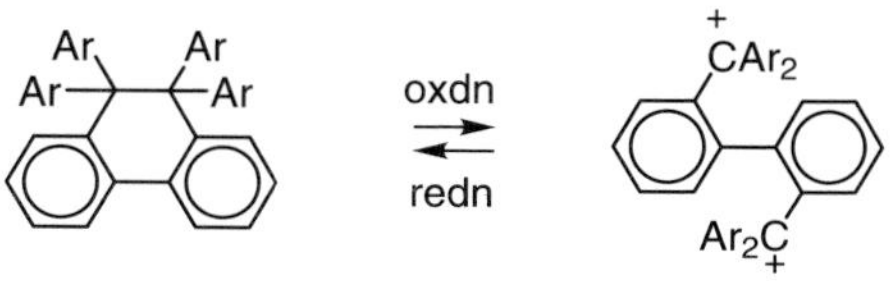

(a: Ar = 4-$Me_2NC_6H_4$; b: Ar = 4-$MeOC_6H_4$)

Fig. 1.21. Dynamic redox properties of hexaphenylethane derivatives

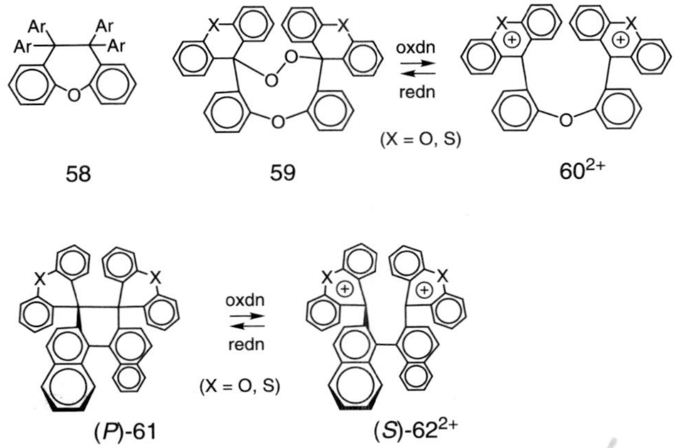

58 59 60^{2+}

(P)-61 (S)-62^{2+}

Fig. 1.22. Dynamic redox properties of bis(triarylmethylenium) derivatives

binaphthylic dications $\mathbf{62}^{2+}$ can be mutually transformed with high yields of the isolated product. Not only the UV–visible spectrum and the molecular geometry but also the circular-dichroism (CD) spectrum were changed drastically for the resolved redox pair (X = S: $\Delta\varepsilon$ +82.0 at 270 nm for (*P*)-**61**; −120 at 288 nm for (*S*)-$\mathbf{62}^{2+}$) [101]. Although there have been only a limited number of examples that exhibit an electrochiroptical response [102,103], much attention will be focused on these systems in the future since they can be considered as multioutput response systems that allow the various methods of readout.

The dispiro skeleton found in **61** is an interesting framework for designing redox systems of this category since the release of strain energy may facilitate the C–C bond fission upon ET. There is another example of dispiro compounds, **63** (Fig. 1.23), that supports this idea. The bis(1,3-dithiole)s **63** are the tricyclic isomers of the cycloalkane-inserted TTFs **32** that give bicyclic dications $\mathbf{33}^{2+}$ by C–C bond making upon 2e-oxidation. The same dications have been obtained from the dispiro compounds **63** (n = 1, 2) upon oxidation accompanied by C–C bond cleavage [77]. Upon 2e-reduction of $\mathbf{33b}^{2+}$ with a bicyclo[4.4.0] skeleton, the dispiro molecule **63b** was isolated in high yield without any by-production of the isomeric diolefin **32b**. In this

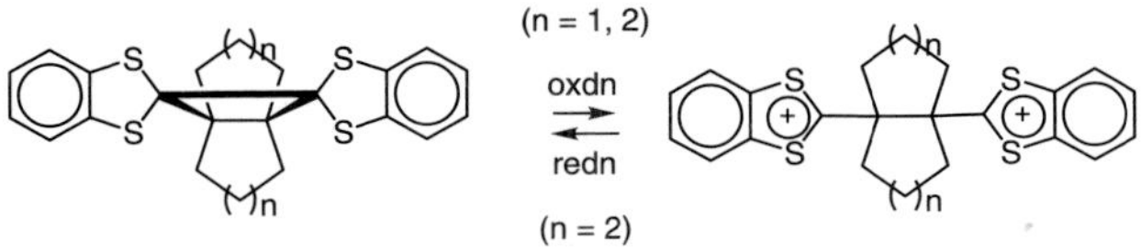

63 (a: n = 1; b: n = 2) 33^{2+}

Fig. 1.23. Dynamic redox properties of a dispiro compound

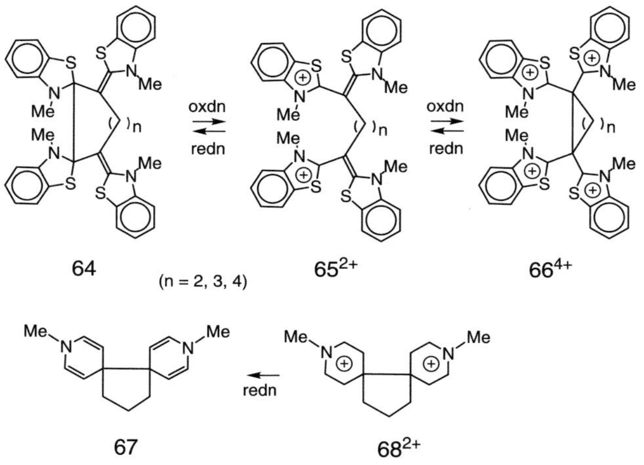

Fig. 1.24. Redox reactions of bis(cyanines)

way, two types of dynamic redox systems (**32a** and **33a**$^{2+}$; **63b** and **33b**$^{2+}$) can be constructed by changing only the number of methylene units of the carbocyclic ring, owing to the divergent reaction pathways of the bicyclic dications **33**$^{2+}$.

Another interesting behavior was observed in the bis(cyanine)-type dications **65**$^{2+}$ (Fig. 1.24), which were transformed into the neutral dispiro compounds **64** upon reduction and into the tetracations **66**$^{4+}$ upon oxidation; each process is accompanied by C–C bond making. The dications **65**$^{2+}$ were regenerated from both **64** and **66**$^{4+}$ by C–C bond fission [104], thus exhibiting two types of dynamic property depending on their oxidation states. Use of cyanine units is advantageous in terms of strong coloration as well as reversibility, compared with the smaller π-systems such as N-methylpyridinium found in the pair of **67** and **68**$^{2+}$ [105].

1.4 Toward the Realization of Unimolecular Memory

This review of dynamic redox systems has shown that they are an interesting class of functionalized dyes showing an optical response to electrochemical input. One of their common features is a facile 2e-transfer at nearly the same potential. This feature favors reversibility of the response, since the intermediary ion radicals tend to undergo side reactions, depending on their kinetic instability. This feature is also advantageous for using dynamic redox systems as unimolecular memories because one molecule can be considered as one digit if the two closed-shell oxidation states are assigned to as 0 and 1 (e.g. the neutral donor is 0 and the dication is 1). A schematic model is illustrated in Fig. 1.25, where the molecules are arranged in an orderly way on the surface, forming a two-dimensional crystalline phase. Although most

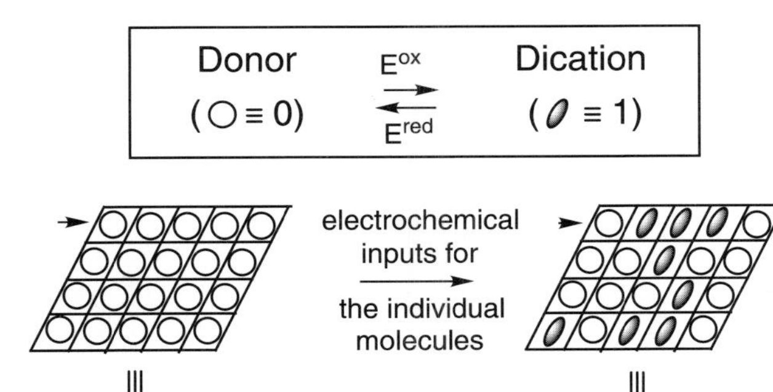

Fig. 1.25. Schematic view of unimolecular memory

of the redox pairs described in this chapter were designed as electrochromic systems, it seems impossible to detect a UV–visible spectral change at the unimolecular level. The second important feature of the present compounds is the drastic structural change induced by electrochemical inputs, which could be detected for a two-dimensional array of molecules by using STM technology [106].

The most advantageous point of the use of dynamic redox systems for realizing unimolecular memory is their electrochemical bistability, especially that of the compounds described in the Sect. 1.3. That is, the E^{ox} values of the donors and the E^{red} values of the dications differ considerably, e.g. by more than 1 V for the pair of **56** and $\mathbf{57}^{2+}$. This corresponds to an energy of about 23 kcal mol^{-1} for electron exchange between **56** and $\mathbf{57}^{2+}$, indicating negligible occurrence at room temperature. In contrast, ordinary redox pairs cannot realize electrochemical memory effects in the unimolecular sense, because exchange between the neutral and oxidized species occurs very rapidly and easily with only a marginal activation energy, as demonstrated by the electrical conduction of p-doped semiconductors. So, the pronounced bistability found in dynamic redox systems is an indispensable feature of a unimolecular-memory material. In conclusion, synthetic chemists working in this area have to be ready to provide promising redox pairs with a detectable structural difference and high bistability and durability before STM technology becomes advanced enough to allow fast electrochemical input to one targeted molecule to be performed freely and to allow very high-speed image collection for readout.

References

1. M.R. Bryce: Adv. Mater. **11**, 11 (1999)
2. F. Wudl, G. Smith, E.J. Hufnage: J. Chem. Soc., Chem. Commun., 1453 (1970)
3. D.S. Acker, R.J. Harder, W.R. Hertler, W. Mahler, L.R. Melby, R.E. Benson, W.E. Mockel: J. Am. Chem. Soc. **82**, 6408 (1960)

4. J. Ferraris, D.O. Cowan, V.V. Walatka, Jr., J.H. Perlstein: J. Am. Chem. Soc. **95**, 948 (1973)
5. J.H. Perlstein: Angew. Chem., Int. Ed. Engl. **15**, 519 (1977)
6. P.M.S. Monk, R.J. Mortimer, D.R. Rosseinsky: *Electrochromism: Fundamentals and Applications*, VCH, Weinheim (1995)
7. S. Hünig, M. Kemmer, H. Wenner, I.F. Perepichka, P. Bäuerle, A. Emge, G. Gescheid: Chem. Eur. J. **5**, 1969 (1999)
8. A.N. Shipway, I. Willner: Acc. Chem. Res. **34**, 421 (2001)
9. Z.J. Donhauser, B.A. Mantooth, K.F. Kelly, L.A. Bumm, J.D. Monnell, J.J. Stapleton, D.W. Price, Jr., A.M. Rawlett, D.L. Allara, J.M. Tour, P.S. Weiss: Science **292**, 2303 (2001)
10. P.L. Boulas, M. Gómez-Kaifer, L. Echegoyen: Angew. Chem., Int. Ed. **37**, 216 (1998)
11. J.-P. Collin, C. Dietrich-Buchecker, P. Gaviña, M.C. Jimenez-Molero, J.-P. Sauvage: Acc. Chem. Res. **34**, 477 (2001)
12. A.R. Pease, J.O. Jeppesen, J.F. Stoddart, Y. Luo, C.P. Collier, J.R. Heath: Acc. Chem. Res. **34**, 433 (2001)
13. S.F. Nelsen, W.C. Hollinsed, C.R. Kessel, J.C. Calabrese: J. Am. Chem. Soc. **100**, 7876 (1978)
14. S.F. Nelsen: Adv. in Electron Transfer Chem. **3**, 167 (1993)
15. S.F. Nelsen, R.F. Ismagilov, K.E. Gentile, D.R. Powell: J. Am. Chem. Soc. **121**, 7108 (1999)
16. S.F. Nelsen, R.F. Ismagilov, D.R. Powell: J. Am. Chem. Soc. **118**, 6313 (1996)
17. T. Suzuki, T. Sakimura, S. Tanaka, Y. Yamashita, H. Shiohara, T. Miyashi: J. Chem. Soc., Chem. Commun. 1431 (1994)
18. K. Lammerstma, M. Marzaghi, G.A. Olah, J.A. Pople, A.J. Kos, P.v.R. Schleyer: J. Am. Chem. Soc. **105**, 5252 (1983)
19. G. Frenking: J. Am. Chem. Soc. **113**, 2476 (1991)
20. H. Bock, K. Ruppert, K. Merzweiler, D. Fenske, H. Goesmann: Angew. Chem. Int. Ed. Engl. **28**, 1684 (1989)
21. N.C. Baenziger, R.E. Buckles, T.D. Simpson: J. Am. Chem. Soc. **89**, 3405 (1967)
22. R. Rathore, S.V. Lindeman, A.S. Kumar, J.K. Kochi: J. Am. Chem. Soc. **120**, 6931 (1998)
23. S. Hünig, M. Kemmer, H. Wenner, F. Barbosa, G. Gescheidt, I.F. Perepichka, P. Bäuerle, A. Emge, K. Peters: Chem. Eur. J. **6**, 2618 (2000)
24. T. Suzuki, H. Shiohara, M. Monobe, T. Sakimura, S. Tanaka, Y. Yamashita, T. Miyashi: Angew. Chem. Int. Ed. Engl. **31**, 455 (1992)
25. H. Shiohara: Master's thesis, Preparation and Properties of Novel Redox Systems by Using the Dithienylmethylene Unit as an Electron-Donating Building Block, Tohoku University (1992)
26. T. Aoshima: Master's thesis, Preparation and Properties of Dendrimers Containing Redox Active Dithienylmethylene Units, Tohoku University (2000)
27. S. Hünig, I.F. Perepichka, M. Kemmer, H. Wenner, P. Bäuerle, A. Emge: Tetrahedron **56**, 4203 (2000)
28. T.M. Bockman, J.K. Kochi: J. Org. Chem. **55**, 4127 (1990)
29. B.A. Scott, S.J. La Place, J.B. Torrance, B.D. Silverman, B. Welber: J. Am. Chem. Soc. **99**, 6631 (1977)

30. P.R. Ashton, V. Balzani, J. Becher, A. Credi, M.C.T. Fyfe, G. Mattersteig, S. Menzer, M.B. Nielsen, F.M. Raymo, J.F. Stoddart, M. Venturi, D.J. Williams: J. Am. Chem. Soc. **121**, 3951 (1999)
31. K. Deuchert, S. Hünig: Angew. Chem., Int. Ed. Engl. **17**, 875 (1978)
32. Z. Yoshida, T. Kawase, H. Awaji, I. Sugimoto, T. Sugimoto, S. Yoneda: Tetrahedron Lett. **38**, 3469 (1983)
33. T. Sugimoto, H. Awaji, I. Sugimoto, Y. Misaki, T. Kawase, S. Yoneda, Z. Yoshida: Chem. Mater. **1**, 535 (1989)
34. Y. Yamashita, M. Tomura, M.B. Zaman, K. Imaeda: Chem. Commun. 1657 (1998)
35. A. Ohta, Y. Yamashita: J. Chem. Soc., Chem. Commun. 1761 (1995)
36. T. Suzuki, T. Yoshino, J. Nishida, M. Ohkita, T. Tsuji: J. Org. Chem. **65**, 5514 (2000)
37. T. Suzuki, S. Mikuni, H. Higuchi, M. Ohkita, T. Tsuji: unpublished results
38. T. Suzuki, H. Higuchi, M. Ohkita, T. Tsuji: Chem. Commun. 1574 (2001)
39. Y. Yamashita, Y. Kobayashi, T. Miyashi: Angew. Chem., Int. Ed. Engl. **28**, 1052 (1989)
40. Y. Yamashita, S. Tanaka, K. Imaeda, H. Inokuchi, M. Sano: J. Org. Chem. **57**, 5517 (1992)
41. Y. Yamashita, K. Ono, S. Tanaka, K. Imaeda, H. Inokuchi: Adv. Mater. **6**, 295 (1994)
42. Y. Yamashita, S. Tanaka, K. Imaeda: Synth. Met. **71**, 1965 (1995)
43. Y. Yamashita, M. Tomura, S. Tanaka, K. Imaeda: Synth. Met. **86**, 1795 (1997)
44. M.R. Bryce, A.J. Moore, M. Hasan, G.J. Ashwell, A.T. Fraser, W. Clegg, M.B. Hursthouse, A.I. Karaulov: Angew. Chem. Int. Ed. Engl. **29**, 1450 (1990)
45. A. Aumüller, S. Hünig: Liebigs Ann. Chem. 618 (1984)
46. U. Schubert, S. Hünig, A. Aumüller: Liebigs Ann. Chem. 1216 (1985)
47. C. Kabuto, Y. Fukazawa, T. Suzuki, Y. Yamashita, T. Miyashi, T. Mukai: Tetrahedron Lett. **27**, 925 (1986)
48. A.M. Kini, D.O. Cowan, F. Gerson, R. Möckel: J. Am. Chem. Soc. **107**, 556 (1985)
49. N. Martín, M. Hanack: J. Chem. Soc., Chem. Commun. 1522 (1988)
50. R. Gómez, J.L. Segura, N. Martín: J. Org. Chem. **65**, 7566 (2000)
51. Y. Yamashita, T. Suzuki, G. Saito, T. Mukai: J. Chem. Soc., Chem. Commun. 1044 (1985)
52. Y. Yamashita, T. Suzuki, G. Saito, T. Mukai: Chem. Lett. 715 (1986)
53. Y. Tsubata, T. Suzuki, Y. Yamashita, T. Mukai, T. Miyashi: Heterocycles **33**, 337 (1992)
54. T. Suzuki, H. Fujii, Y. Yamashita, C. Kabuto, S. Tanaka, M. Harasawa, T. Mukai, T. Miyashi: J. Am. Chem. Soc. **114**, 3034 (1992)
55. P.W. Kenny: J. Chem. Soc., Perkin Trans. 2 907 (1995)
56. S. Yamaguchi, T. Hanafusa, T. Tanaka, M. Sawada, K. Kondo, M. Irie, H. Tatemitsu, Y. Sakata, S. Misumi: Tetrahedron Lett. **21**, 2411 (1986)
57. H. Meyer: Monatsh. Chem. **30**, 165 (1909)
58. E. Harni, G.M.J. Schmidt: J. Chem. Soc. 3295 (1954)
59. R. Korenstein, K.A. Muszkat, S. Sharafy-Ozeri: J. Am. Chem. Soc. **95**, 6177 (1973)
60. T. Suzuki, T. Fukushima, T. Miyashi, T. Tsuji: Angew. Chem. Int. Ed. Engl. **36**, 2495 (1997)

61. D.H. Evans, R.W. Busch: J. Am. Chem. Soc. **104**, 5057 (1982)
62. L.N. Ferguson, J.C. Nnadi: J. Chem. Edu. **42**, 529 (1965)
63. R. Gleiter, W. Schäfer: Acc. Chem. Res. **23**, 369 (1990)
64. K.-D. Asmus: Acc. Chem. Res. **12**, 436 (1979)
65. R.W. Alder, N.C. Goode: J. Chem. Soc., Chem. Commun. 173 (1976)
66. R.W. Alder, R. Gill, N.C. Goode: J. Chem. Soc., Chem. Commun. 973 (1976)
67. R.W. Alder, R.B. Sessions: J. Am. Chem. Soc. **101**, 3651 (1979)
68. W.K. Musker: Acc. Chem. Res. **13**, 200 (1980)
69. F. Iwasaki, N. Toyoda, R. Akaishi, H. Fujihira, N. Furukawa: Bull. Chem. Soc. Jpn. **61**, 2563 (1988)
70. N. Furukawa: Bull. Chem. Soc. Jpn **70**, 2571 (1997)
71. N. Furukawa, K. Kobayashi, S. Sato: J. Organomet. Chem. **611**, 116 (2000)
72. M. Horner, S. Hünig: J. Am. Chem. Soc. **99**, 6122 (1977)
73. M. Horner, S. Hünig: Liebigs Ann. Chem. 69 (1983)
74. K. Hesse, S. Hünig: Liebigs Ann. Chem. 740 (1985)
75. W. Freund, S. Hünig: Helv. Chem. Acta **70**, 929 (1987)
76. W. Freund, S. Hünig: J. Org. Chem. **52**, 2154 (1987)
77. T. Suzuki, M. Kondo, T. Nakamura, T. Fukushima, T. Miyashi: Chem. Commun. 2325 (1997)
78. T. Suzuki, T. Yoshino, M. Ohkita, T. Tsuji: J. Chem. Soc., Perkin Trans. 1 3417 (2000)
79. G.A. Olah, J.S. Staral, L.A. Paquette: J. Am. Chem. Soc. **98**, 1267 (1976)
80. R. Rathore, P. Le Magueres, S.V. Lindeman, J.K. Kochi: Angew. Chem., Int. Ed. **39**, 809 (2000)
81. T. Suzuki, H. Higuchi, M. Ohkita, M. Tsuji: Chem. Commun. 1574 (2001)
82. T. Suzuki, H. Takahashi, J. Nishida, T. Tsuji: Chem. Commun. 1331 (1998)
83. H. Kurata, T. Shimoyama, K. Matsumoto, T. Kawase, M. Oda: Bull. Chem. Soc. Jpn. **74**, 1327 (2001)
84. M. Rosi, A. Sgamellotti, F. Franceschi, C. Floriani: Chem. Eur. J. **5**, 2914 (1999)
85. F. Franceschi, E. Solari, R. Scopelliti, C. Floriani: Angew. Chem., Int. Ed. **39**, 1685 (2000)
86. C. Fritze, G. Erker, R. Fröhlich: J. Organomet. Chem. **501**, 41 (1995)
87. Y. Murata, K. Komatsu, T.S.M. Wan: Tetrahedron Lett. **39**, 7061 (1996)
88. Y. Murata, M. Shiro, K. Komatsu: J. Am. Chem. Soc. **119**, 8117 (1997)
89. P. Maslak, J.N. Narvaez, T.M. Vallombroso, Jr., B.A. Watson: J. Am. Chem. Soc. **117**, 12380 (1995)
90. G. Dyker, J. Körning, P. Bubenitschek, P.G. Jones: Liebigs Ann./Recueil 203 (1997)
91. W.D. Hounshell, D.A. Dougherty, J.P. Hummel, K. Mislow: J. Am. Chem. Soc. **99**, 1916 (1977)
92. K.K. Baldridge, T.R. Batterby, R.V. Clark, J.S. Siegel: J. Am. Chem. Soc. **119**, 7048 (1997)
93. T. Suzuki, K. Ono, J. Nishida, H. Takahashi, T. Tsuji: J. Org. Chem. **65**, 4944 (2000)
94. J.M. McBride: Tetrahedron **30**, 2009 (1974)
95. M. Stein, W. Winter, A. Rieker: Angew. Chem., Int. Ed. Engl. **17**, 692 (1978)
96. B. Kahr, D.V. Engen, K. Mislow: J. Am. Chem. Soc. **108**, 8305 (1986)
97. G. Wittig, H. Petri: Justus Liebigs Ann. Chem. **505**, 17 (1933)

98. T. Suzuki, J. Nishida, T. Tsuji: Angew. Chem. Int. Ed. Engl. **36**, 1329 (1997)
99. T. Suzuki, J. Nishida, T. Tsuji: Chem. Commun. 2193 (1998)
100. T. Suzuki, J. Nishida, M. Ohkita, T. Tsuji: Angew. Chem., Int. Ed. **39**, 1804 (2000)
101. T. Suzuki, J. Nishida, M. Ohkita, T. Tsuji: Angew. Chem., Int. Ed. **40**, 3251 (2001)
102. C. Westermeier, H.-C. Gallmeier, M. Komma, J. Daub: Chem. Commun. 2427 (1999)
103. G. Beer, C. Niederalt, S. Grimme, J. Daub: Angew. Chem., Int. Ed. **39**, 3252 (2001)
104. S. Hünig, C.A. Briehn, P. Bäuerle, A. Emge: Chem. Eur. J. **7**, 2745 (2001)
105. T. Muramatsu, A. Toyota, M. Kudou, Y. Ikegami, M. Watanabe: J. Org. Chem. **64**, 7249 (1999)
106. B. Ohtani, A. Shintani, K. Uosaki: J. Am. Chem. Soc. **121**, 6515 (1999)

2 Photoswitching of Intramolecular Magnetic Interaction Using Photochromic Compounds

Kenji Matsuda, Masahiro Irie

Summary. Recent advances in organic material chemistry have helped to realize several novel organic functional materials, including organic conductors, magnets, and nonlinear optical materials. One of the current frontiers in this field is molecular electronics, where a single molecule plays the role of one component in an electric circuit. For the realization of this goal, highly integrated materials with multiple functions are necessary. Photochromism is the light-induced reversible transformation of chemical species between two isomers that have different absorption spectra. The two isomers differ from each other not only in the absorption spectra but also in various physical and chemical properties. These property changes can, in principle, be utilized to control the function of organic materials. Two organic radicals placed at the two edges of a π-conjugative molecule interact magnetically through the framework via the exchange interaction J. Therefore photocontrol of the magnetism of the system is possible when the photochromic moiety is used as a backbone. In this chapter we will describe studies of the photoswitching of the intramolecular magnetism by incorporating two radical moieties into a photochromic diarylethene spin coupler. Photoswitching using a diarylethene dimer, which showed similar electric-circuit behavior, will also be discussed.

2.1 Introduction

Nanomolecular systems, as a step towards the realization of molecular electronic devices, are considered important, challenging targets in an interdisciplinary area between chemistry and physics [1,2]. For the realization of molecular electronic devices, various kinds of molecular parts are necessary: wires, memories, switches, rectifiers, and so on. Several organic compounds which have unique functions, such as photochromism, conductivity, magnetism, and nonlinear optical properties, have been proposed for these parts [3]. A methodology to assemble such basic compounds into working devices is crucial for the development of practical applications.

In this chapter, we shall describe the synthesis and evaluation of diarylethene photochromic systems bearing nitronyl nitroxide radicals. These systems are hybrids of photochromic molecules and magnetic molecules. Since the intramolecular magnetic interaction is switched by photoirradiation, these systems behave like logic circuits in principle (Fig. 2.1).

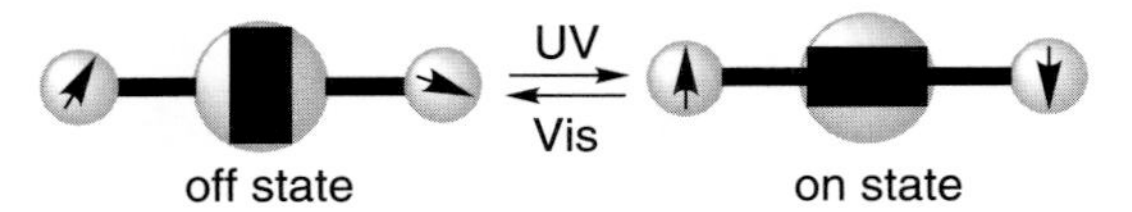

Fig. 2.1. Schematic drawing of photoswitching of intramolecular magnetic interaction

2.2 Photochromic Compounds

A photochromic reaction is an isomerization reaction induced by photoirradiation [4–6]. Upon photoirradiation, photochromic compounds reversibly change their molecular properties, such as their absorption and fluorescence spectra, refractive indices, geometrical structures, dielectric constants, oxidation/reduction potentials, and chiroptical properties. These property changes have been widely used to switch the physical and chemical functions of molecular systems containing photochromic units. In these photoresponsive systems, photoirradiation can be considered as an input signal, and the molecular property changes can be considered as an output signal.

Some typical photochromic compounds are listed in Fig. 2.2. Azobenzenes have been widely used as a component of photoresponsive units utilizing photochemically induced *cis–trans* conformational motion [7–13]. The changes in geometrical structure and polarity are used to produce a mechanichal effect. Thioindigos [14] and spiropyrans [15] have also been used in several photoswitching materials.

Although various types of photochromic compounds have been adopted as switching units, fulgides and diarylethenes have a unique feature; both isomers of those compounds are thermally stable. In other words, the switching reactions are induced only by photoirradiation and not by heat. This feature is ideal for the application of these molecules in memories and switches. In particular, diarylethenes are the most promising compounds for molecular-scale memories and switch devices [16–21], because they exhibit excellent photochromic performance. Both isomers are thermally stable even at 100 °C, and they exhibit high fatigue resistance ($> 10^4$ coloration/decoloration cycles) and very rapid response (~ 1 ps).

The open-ring and closed-ring isomers of the diarylethenes have substantially different electronic and geometrical structures. The most striking difference is that while the π-systems of the two aryl rings are discontinuous in the open-ring isomer, the two π-systems are connected in the closed-ring isomer. Details of this key feature will be discussed below.

2.3 Intramolecular Magnetic Interaction

In contrast to typical organic molecules, which have closed-shell electronic structures, free radicals have one or more unpaired electrons. When two

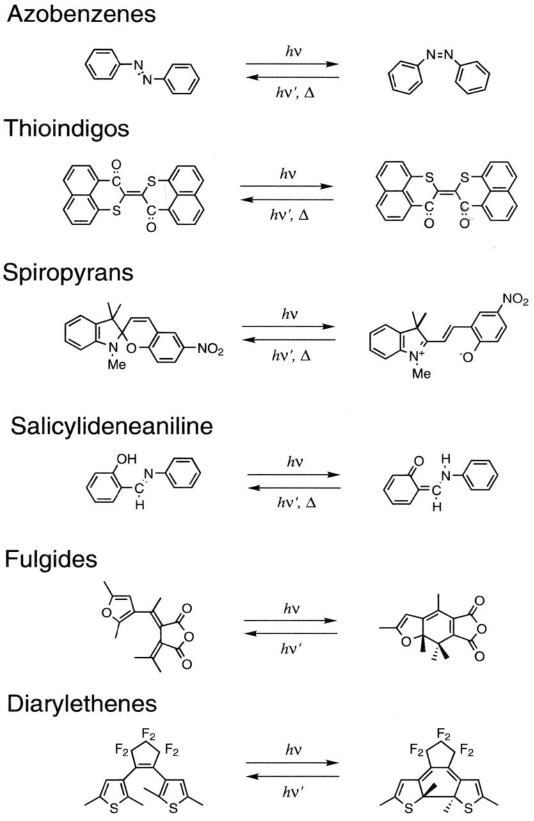

Azobenzenes

Thioindigos

Spiropyrans

Salicylideneaniline

Fulgides

Diarylethenes

Fig. 2.2. Examples of photochromic compounds

unpaired electrons are placed at the two ends of a π-conjugated chain, the two spins of the unpaired electrons interact magnetically [22]. This magnetic interaction depends on the nature of the π-conjugated spacers. The direction of the spin alignment, or the sign of the exchange interaction, is regulated by the Ovchinnikov rule [23] and the magnitude becomes weaker with an increase in the spacer length. The magnetochemistry of such biradicals has been intensively investigated and great effort has been devoted to synthesizing organic ferromagnetic materials [24–29].

We note here the classification of biradicals (Fig. 2.3). If a biradical has no resonant closed-shell structure, the biradical is classified as a non-Kekulé biradical. If a biradical has a resonant closed-shell structure, the closed-shell structure is more stable, therefore, and the molecule exists as a normal closed-shell Kekulé molecule. The non-Kekulé biradicals can be further classified as disjoint and nondisjoint biradicals [30]. Trimethylenemethane (**1**) is a typical example of a nondisjoint biradical. The two singly occupied molecular orbitals

(SOMOs) of **1** overlap in space. In this case, the intramolecular magnetic interaction between the radicals is ferromagnetic and strong. On the other hand, tetramethyleneethane (**2**) is a disjoint biradical. The two SOMOs of **2** do not overlap in space. The intramolecular magnetic interaction between the radicals is very weak; the singlet and triplet states of the biradical are nearly degenerate [31]. Butadiene (**3′**) is an example of a normal Kekulé molecule and **3′** has a resonant biradical structure, 2-butene-1,4-diyl (**3**). In this case, the closed-shell structure **3′** is the ground state, so the ground electronic state has no unpaired electrons. In other words, the magnetic interaction between the two spins in the structure **3** is strongly antiferromagnetic.

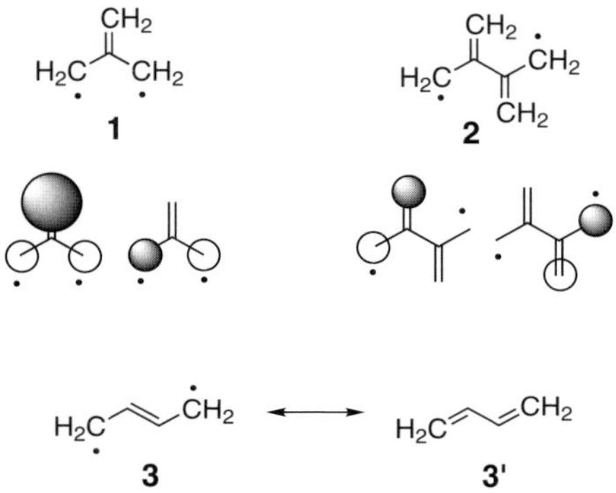

Fig. 2.3. Molecular formula of trimethylenemethane (**1**), tetramethyleneethane (**2**), and butadiene (**3**), with the shapes of the two SOMOs (singly occupied molecular orbitals) of **1** and **2**

If the nature of the π-conjugated spacer can be switched using a photochromic spin coupler, the magnetic interaction can be controlled by photoirradiation. The work described below is based on this idea. We found diarylethenes to be very effective spacers because the π-conjugated chain length can effectively be switched.

2.4 Photoswitching Using a Single Photochromic Molecule

2.4.1 Molecular Design

There is a characteristic feature in the electronic structural changes of diarylethenes. Figure 2.4 shows the open-ring isomer **4a** and closed-ring isomer **4b** of a radical-substituted diarylethene, together with their simplified structures. While there is no resonant closed-shell structure for **4a**, there exists **4b′** as the resonant quinoid-type closed-shell structure for **4b**. **4a** is a non-Kekulé biradical, and **4b** is a normal Kekulé molecule. In other words, **4a** has two unpaired electrons, but **4b** has no unpaired electrons. The calculated shapes of the two SOMOs of **4a** are separated in the molecule and there is no overlap [32]. This configuration is a typical disjoint biradical, in which the intramolecular radical–radical interaction is weak. In the open-ring isomer, the bond alternation is discontinued at the 3-position of the thiophene rings. This is the origin of the disjoint nature of the electronic configuration of **4a**. **4a** is a disjoint non-Kekulé biradical and corresponds to tetramethyleneethane, **2**, in the former example.

However, the closed-ring isomer **4b′** is a normal Kekulé molecule. In this case, the ground electronic state has no unpaired electrons. In this singlet ground state, the magnetic interaction is strongly antiferromagnetic. **4b′** corresponds to butadiene, **3′**, in the former example.

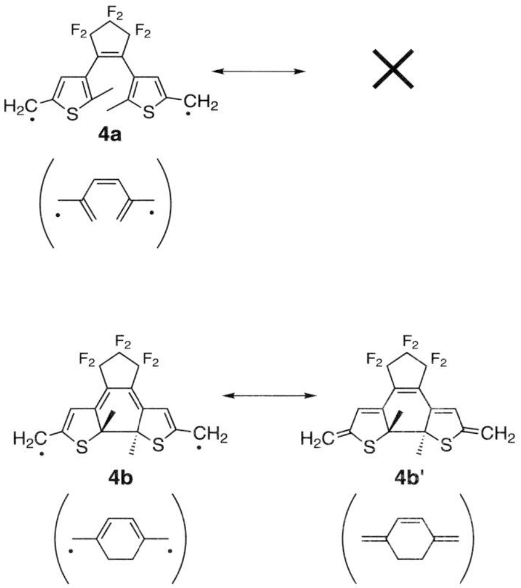

Fig. 2.4. Resonance structures of open- and closed-ring isomers of **4a** and **4b**. The parentheses contain simplified structures

The electronic structural change of a radical-substituted diarylethene accompanying photoisomerization is a transformation of a disjoint non-Kekulé structure to a closed-shell Kekulé structure. It can be inferred from the above consideration that the interaction between spins in the open-ring isomer of diarylethene is weak, while significant antiferromagnetic interaction takes place in the closed-ring isomer. In other words, the open-ring isomer is in the "OFF" state and the closed-ring isomer is in the "ON" state.

2.4.2 Switching of Magnetic Interaction Detected by Susceptibility Measurement

We designed and synthesized molecule **5a** (Fig. 2.5) by choosing 1,2-bis(2-methyl-1-benzothiophen-3-yl)perfluorocyclopentene as a photochromic spin coupler and nitronyl nitroxides as spin sources [32–34]. Crystalline **5a** was obtained as dark blue plates.

In solution, **5a** showed ideal photochromic behavior on irradiation with UV and visible light (Fig. 2.6). Although the radical moiety absorbs in the region from 550 to 700 nm, this did not prevent the photochromic reaction. Almost 100% photochemical conversion was observed in both the cyclization from the open-ring isomer **5a** to the closed-ring isomer **5b** and the cycloreversion from **5b** to **5a**. For the practical use of photochromic devices, high conversion is one of the most important characteristics.

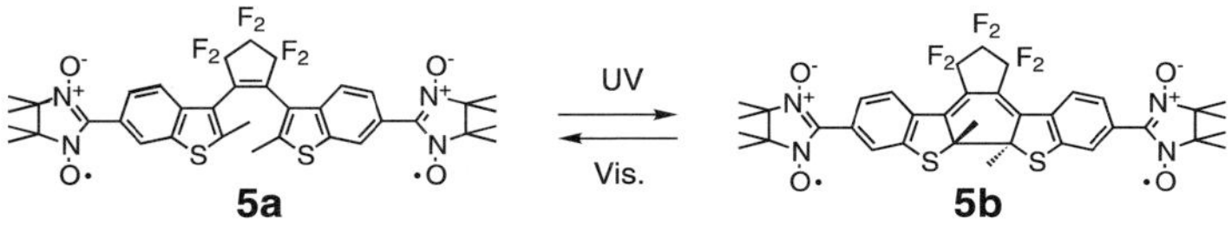

Fig. 2.5. Photochromic reaction of **5**

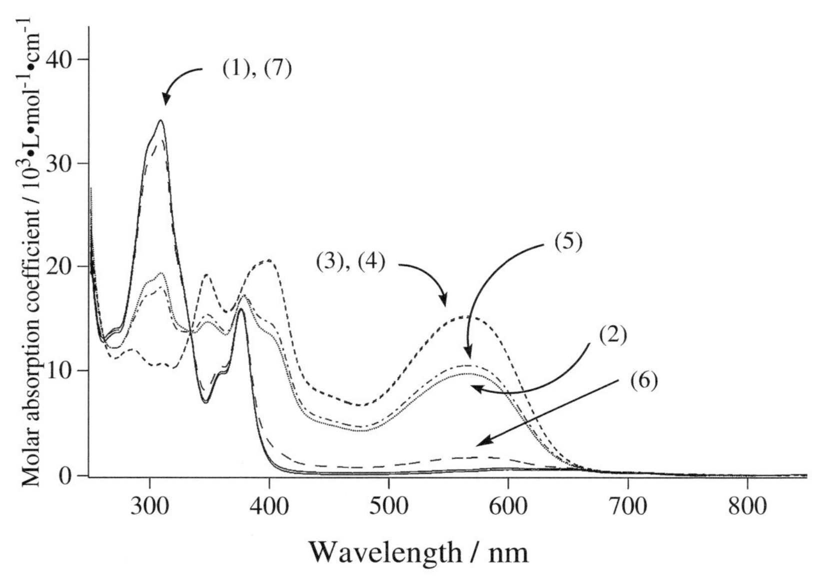

Fig. 2.6. Changes in absorption spectrum of **5a** associated with photochromism (ethyl acetate solution, 1.7×10^{-5} M): (1) initial; (2) irradiation with 313 nm light for 1 min; (3) 5 min; (4) 10 min; (5) irradiation with 578 nm light for 5 min; (6) 30 min; (7) 60 min (from K. Matsuda et al., Chem. Lett. 16 (2000), reprinted with permission)

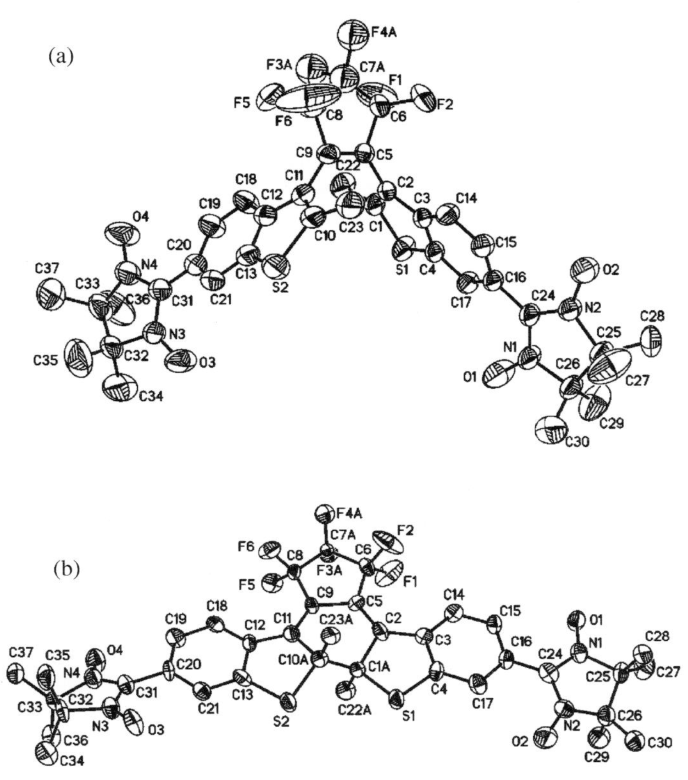

Fig. 2.7. ORTEP drawings of (a) open-ring isomer **5a** and (b) closed ring isomer **5b** (from K. Matsuda et al., J. Am. Chem. Soc. **122**, 7195 (2000), reprinted with permission)

The crystal structures of both isomers, **5a** and **5b**, were determined by X-ray crystallographic analysis. Fig. 2.7 shows ORTEP drawings of the open-ring isomer **5a** and the closed-ring isomer **5b**. While the dihedral angle between the benzothiophene ring and the perfluorocyclopentene ring in the open-ring isomer **5a** was 86.1°, that in the closed-ring isomer **5b** was 2.6°. The change in geometrical structure associated with the photochromism is rather large, which is favorable for the switching of the interaction.

The magnetic susceptibilities of **5a** and **5b** were measured with a SQUID susceptometer on samples in microcrystalline form. χT–T plots are shown in Fig. 2.8. The data were analyzed in terms of a modified singlet–triplet two-spin model (the Bleaney–Bowers type) in which two spins ($S = 1/2$) are coupled antiferromagnetically within a biradical molecule by an exchange interaction J [35]. The best-fit parameters obtained by means of a least-squares method were $2J/k_\mathrm{B} = -2.2 \pm 0.04\,\mathrm{K}$ for **5a** and $2J/k_\mathrm{B} = -11.6 \pm 0.4\,\mathrm{K}$ for **5b**. Although the interaction ($2J/k_\mathrm{B} = -2.2\,\mathrm{K}$) between the two spins in the open-ring isomer **5a** is small, the spins of **5b** have a remarkable antiferromagnetic interaction ($2J/k_B = -11.6\,K$).

The open-ring isomer **5a** has a twisted molecular structure and a disjoint electronic configuration. On the other hand, the closed-ring isomer **5b** has a planar molecular structure and a nondisjoint electronic configuration. The photoinduced change in magnetism agrees well with the prediction that the open-ring isomer is in an "OFF" state and the closed-ring isomer is in an "ON" state.

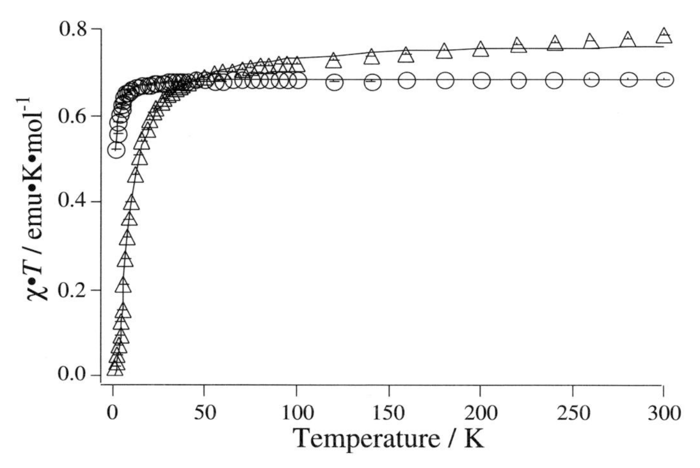

Fig. 2.8. χT–T plot of biradicals **5a** (○) and **5b** (△) (from K. Matsuda et al., Chem. Lett. 16 (2000), reprinted with permission)

2.4.3 Switching of Magnetic Interaction Detected by ESR Spectra

Although the switching of the exchange interaction was detected by a susceptibility measurement of the biradical **5**, both the open- and the closed-ring isomers **5a** and **5b** had 9-line ESR spectra because the exchange interaction between the two radicals was much stronger than the hyperfine coupling constant in both isomers. For it to be possible to detect the change of the exchange interaction by ESR spectroscopy, the value of the interaction should be comparable to the hyperfine coupling constant. Therefore we have designed and synthesized the biradicals **6a** and **7a**, in which *p*-phenylene spacers are introduced to control the strength of the exchange interaction (Fig. 2.9) [36,37].

Nitronyl nitroxides themselves have two identical nitrogen atoms, so that they give 5-line ESR spectra with relative intensities 1:2:3:2:1 and a 7.5 G spacing. When two nitronyl nitroxides are magnetically coupled via an exchange interaction, the biradical gives a 9-line ESR spectrum with relative

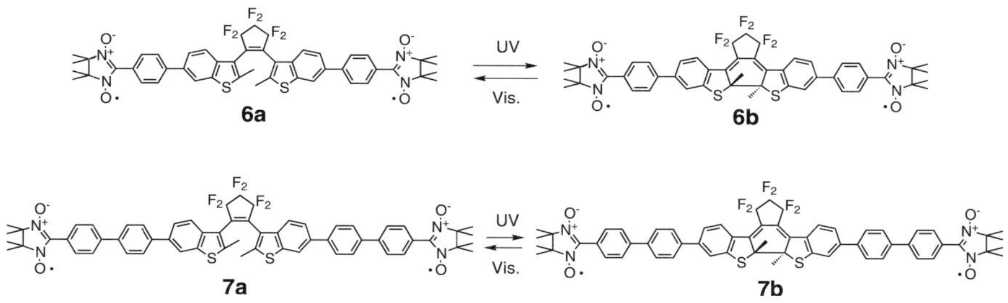

Fig. 2.9. Photochromic reactions of **6** and **7**

intensities 1:4:10:16:19:16:10:4:1 and a 3.7 G spacing. If the exchange interaction is smaller than the hyperfine coupling in the biradical, the two nitroxide radicals are magnetically independent and give the same spectrum as for the independent monoradical. In intermediate situations the spectrum becomes complex [38,39].

The Diarylethenes **6a** and **7a** underwent reversible photochromic reactions in ethyl acetate solution when subjected to alternate irradiation with 313 nm UV light and 578 nm visible light (Fig. 2.10). The visible absorption maxima of the closed-ring isomers of the diarylethenes **5**, **6**, and **7** showed a hypsochromic shift with an increase in the chain length. This unusual phenomenon is attributed to the smaller contribution of the resonant quinoid structure as the chain length becomes longer.

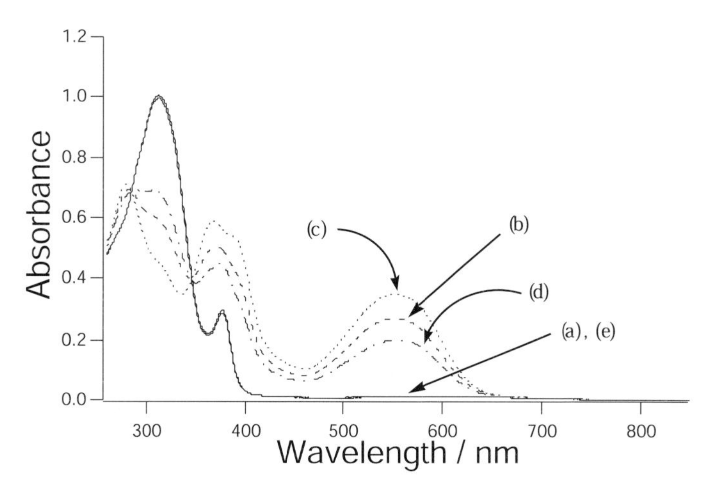

Fig. 2.10. Changes in absorption spectrum of **6a** associated with photochromism (ethyl acetate solution, 1.3×10^{-5} M): (a) initial; (b) irradiation with 313 nm light for 2 min; (c) 10 min; (d) irradiation with 578 nm light for 2 min; (e) 20 min; (f) 60 min (from K. Matsuda et al., J. Am. Chem. Soc. **122**, 8309 (2000), reprinted with permission)

The quantum yields of the cyclization and cycloreversion reactions of **5**, **6**, and **7** were measured. The quantum yields of both the cyclization and cycloreversion reactions were lower than those for the normal diarylethene. The quantum yield of the cyclization reactions increased from 0.040 in **5a** to 0.13 in **7a** upon increasing the π-conjugated chain length. Energy transfer from the central diarylethene to the radical moiety is believed to reduce the cyclization quantum yield of **5a**. More interestingly, it was shown that the cycloreversion quantum yield increased from 0.0010 in **5b** to 0.062 in **7b** upon increasing the π-conjugated chain length. In this case, besides the effect of the energy transfer, the contribution of the resonant quinoid nature to the structure of **5b** is believed to suppress the cycloreversion reaction.

The changes in the ESR spectra accompanying the photochromic reaction were examined for the diarylethenes **6** and **7**. Figure 2.11 shows the ESR spectra at different stages of the photochromic reaction of **6**. The ESR spectrum of **6a** showed a complex of 15 lines. This suggests that the two spins of the nitronyl nitroxide radicals are coupled by an exchange interaction that is comparable to the hyperfine coupling constant. Upon irradiation with 366 nm light, the spectrum converted completely to a 9-line spectrum corresponding to the closed-ring isomer **6b**. The 9-line spectrum indicates that the exchange interaction between the two spins in **6b** is much larger than the hyperfine coupling constant.

A change in the ESR spectra was also observed for **7**. The open-ring isomer **7a** had a 5-line spectrum, while the closed-ring isomer **7b** had a distorted 9-line spectrum. A simulation of the ESR spectra was performed. The

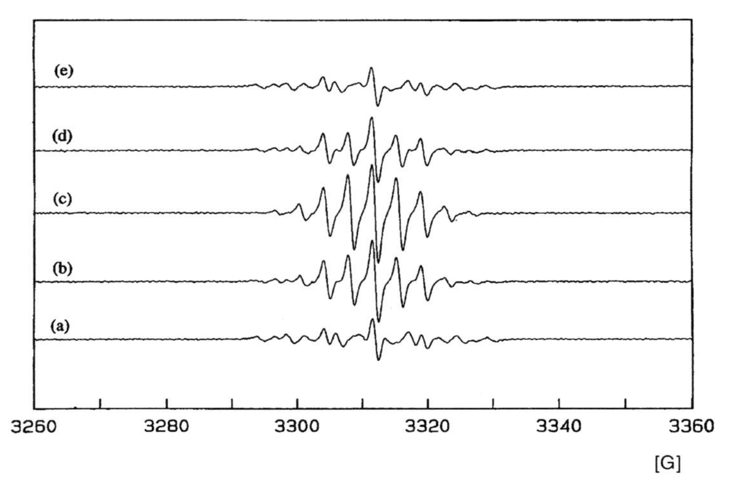

Fig. 2.11. Changes in the ESR spectra of **6a** associated with photochromism (benzene solution, 1.1×10^{-4} M): (a) initial; (b) irradiation with 366 nm light for 1 min; (c) 4 min; (d) irradiation with > 520 nm light for 20 min; (e) 50 min (from K. Matsuda et al., J. Am. Chem. Soc. **122**, 8309 (2000), reprinted with permission)

Table 2.1. Difference in exchange interaction between open- and closed-ring isomers of biradicals **5**–**9**. The values were estimated by fitting of the susceptibility data (**5**) or simulation of the ESR spectra (**6**–**9**)

	Open-ring isomer		Closed-ring isomer	
	ESR line shape	$\lvert 2J/k_B K\rvert$	ESR line shape	$\lvert 2J/k_B K\rvert$
5	9 lines	2.2	9 lines	11.6
6	15 lines	$\begin{cases} 1.2 \times 10^{-3} \\ < 3 \times 10^{-4} \end{cases}$	9 lines	> 0.04
7	5 lines	$< 3 \times 10^{-4}$	Distorted 9 lines	0.010
8	13 lines	$\begin{cases} 5.6 \times 10^{-3} \\ < 3 \times 10^{-4} \end{cases}$	9 lines	> 0.04
9	5 lines	$< 3 \times 10^{-4}$	9 lines	> 0.04

exchange interaction decreases with an increase in the π-conjugated chain length, as shown in Table 2.1. The change in the exchange interaction in **7** upon photoirradiation was more than 30-fold. This result shows a very large switching effect in diarylethenes and suggests the superiority of diarylethenes as molecular switching units. Although the absolute value of the exchange interaction is small, information about the spins can clearly be transmitted through the closed-ring isomer, and the switching can be detected by ESR spectroscopy.

Oligothiophenes are good candidates for conductive molecular wires [40,41]. The thiophene-2,5-diyl moiety has been used as a molecular wire unit for energy and electron transfer and serves as a stronger magnetic coupler than *p*-phenylene [42]. Therefore, we have synthesized then diarylethenes **8a** and **9a**, which have two nitronyl nitroxide radicals at both ends of a molecule containing oligothiophene spacers, and studied their photo- and magnetochemical properties (Fig. 2.12) [43,44].

Photochromic reactions and a change in the ESR spectra were also observed for **8a** and **9a**. Table 2.1 lists the exchange interactions between the two diarylethene-bridged nitronyl nitroxide radicals. For all five biradicals, the closed-ring isomers have stronger interactions than do the open-ring iso-

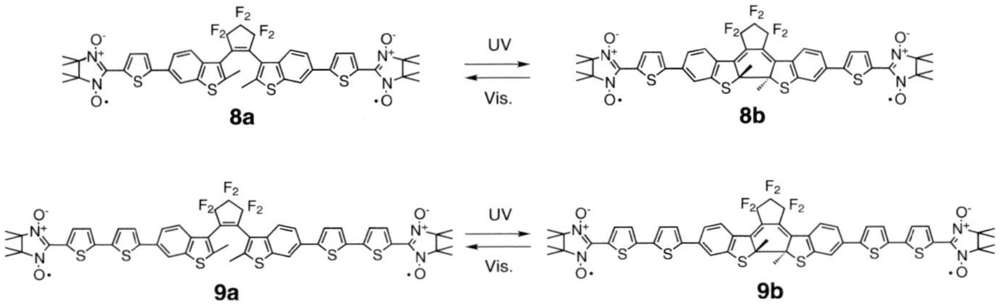

Fig. 2.12. Photochromic reactions of **8** and **9**

mers. The exchange interactions through oligothiophene spacers were larger than in the corresponding biradicals with oligophenylene spacers. The efficient π-conjugation in thiophene spacers resulted in large exchange interactions between the two nitronyl nitroxide radicals. In the case of bithiophene spacers, the difference in the exchange interactions between open- and closed-ring isomers was estimated to be more than 150-fold.

2.5 Photoswitching Using an Array of Photochromic Molecules

In the previous section we have demonstrated that the exchange interaction between two nitronyl nitroxide radicals located at the two ends of a diarylethene could be photoswitched reversibly by alternate irradiation with ultraviolet and visible light. The difference in the exchange interactions between the two switching states was more than 150-fold. It was demonstrated that ESR spectra can be used as a good tool for detecting small changes in magnetic interaction in these molecular systems. In this section, photoswitching of the intramolecular magnetic interaction using a diarylethene dimer will be presented [45].

When a diarylethene dimer is used as a switching unit, there are three kinds of photochromic states; open–open (OO), closed–open (CO), and closed–closed (CC). From the analogy with an electric circuit, it can be inferred that the dimer has two switching units in series. We have designed the dimer **10**, which has 28 carbon atoms between two nitronyl nitroxide radicals (Fig. 2.13). When the two radicals are separated by the 28 conjugated carbon

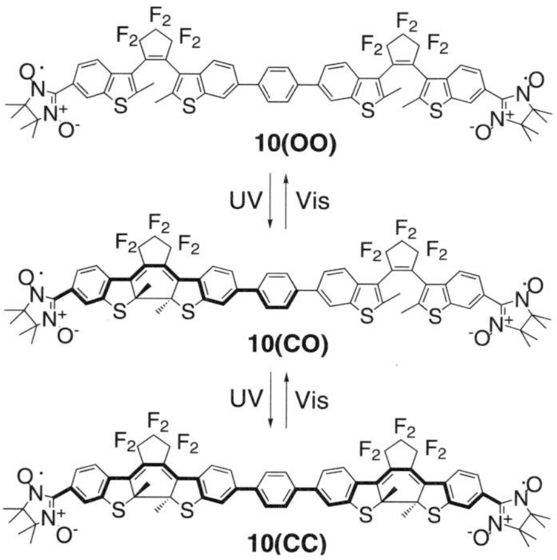

Fig. 2.13. Photochromic reaction of **10**. The bold lines indicate the connectivity of the bond alternation

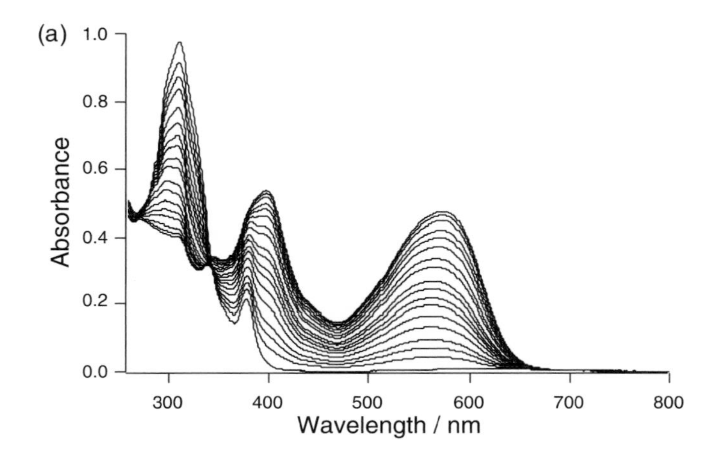

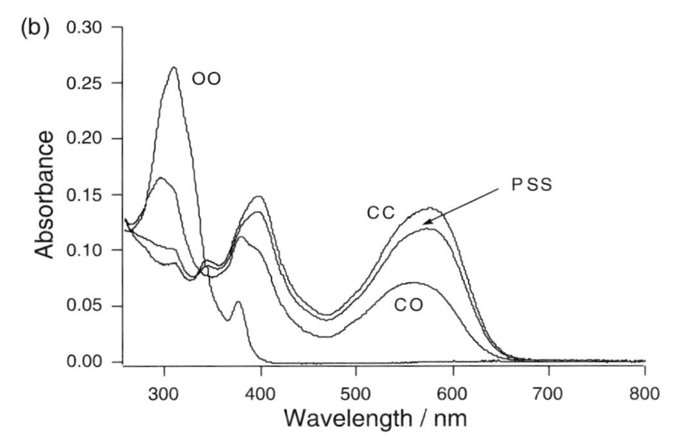

Fig. 2.14. (**a**) Changes in the absorption spectrum of **10** under irradiation with 313 nm light. Initial, 20 s, 40 s, 60 s, 2 min, 3 min, 4 min, 5 min, 7 min, 9 min, 15 min, 20 min, 30 min, 40 min, 50 min, 70 min, 90 min, 120 min. (**b**) Absorption spectra of **10(OO)**, **10(CO)**, **10(CC)** and in the photostationary state under irradiation with 313 nm light (from K. Matsuda et al., J. Am. Chem. Soc. **123**, 9896 (2001), reprinted with permission)

atoms, the 5-line and 9-line spectra are clearly distinguishable upon irradiation. A p-phenylene spacer was introduced so that a cyclization reaction could occur at both diarylethene moieties. The bond alternation is interrupted at the open-ring-form moieties of **10(OO)** and **10(CO)**. As a result, the two spins at the ends of **10(OO)** and **10(CO)** cannot interact with each other. On the other hand, the π-system of **10(CC)** is delocalized throughout the molecule and an exchange interaction between the two radicals is expected to occur.

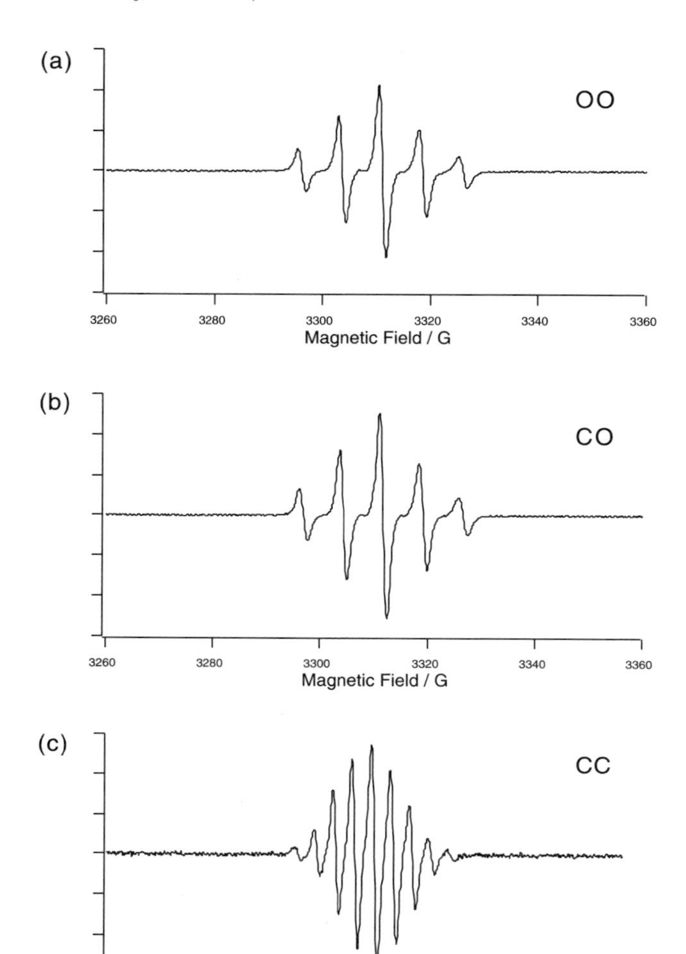

Fig. 2.15. X-band ESR spectra measured at room temperature in benzene (9.32 GHz): (**a**) **10(OO)**, (**b**) **10(CO)**, and (**c**) **10(CC)** (from K. Matsuda et al., J. Am. Chem. Soc. **123**, 9896 (2001), reprinted with permission)

The molecule **10(OO)** underwent photochromic reaction when subjected to alternate irradiation with UV and visible light. Upon irradiation of an ethyl acetate solution of **10(OO)** with 313 nm light, an absorption at 560 nm appeared (Fig. 2.14a). This absorption grew and shifted and the system reached a photostationary state after 120 min. The color of the solution changed from pale blue to red–purple, and then to blue–purple. The red spectral shift suggests the formation of **10(CC)**. The isosbestic point was maintained at an initial stage of irradiation, but it later deviated. The blue-purple solution was completely bleached by irradiation with 578 nm light. **10(CO)** and **10(CC)** were isolated from the blue-purple solution by HPLC (high-performance liquid chromatography). The spectra of **10(OO)**, **10(CO)**, and **10(CC)** are shown in Fig. 2.14b along with the spectrum in the photostationary state under irradiation with 313 nm light. **10(CC)** has an absorption maximum at 576 nm, which is red-shifted by as much as 16 nm in comparison with **10(CO)**.

ESR spectra of isolated **10(OO)**, **10(CO)**, and **10(CC)** were measured in benzene at room temperature (Fig. 2.15). The spectra of **10(OO)** and **10(CO)** have 5 lines, suggesting that the exchange interaction between the two nitronyl nitroxide radicals is much smaller than the hyperfine coupling constant ($|2J/k_B| < 3 \times 10^{-3}$ K). However, the spectrum of **10(CC)** clearly has 9 lines, indicating that the exchange interaction between the two spins is much larger than the hyperfine coupling constant ($|2J/k_B| > 0.04$ K). The result indicates that each diarylethene chromophore serves as a switching unit that can control the magnetic interaction: the magnetic interaction between the terminal nitronyl nitroxide radicals was controlled by the switching units in series.

2.6 Concluding Remarks

In this chapter we have demonstrated the photoswitching of an intramolecular magnetic interaction using photochromic diarylethenes. Not only a monomeric diarylethene but also an array of diarylethenes were proved to work as photoswitches. In this system, the input signal is photoirradiation and the output signal is the exchange interaction. The strength of this exchange interaction can be read out by the shape of the ESR signal. The assembly of such a switch into more realistic logic circuits remains a problem to be solved.

Acknowledgments

We thank Professor Paul M. Lahti and Professor B. Kriste for providing the program BIRADG. We thank Mr. Mitsuyoshi Matsuo for his contribution to this project. We thank Dr. Thomas S. Hughes for reviewing the manuscript. This work was supported by the CREST program of Japan Science and Technology Corporation and by a Grant-in Aid for Scientific Research on Priority Area "Creation of Delocalized Electronic Systems" from the Ministry of Education, Science, Culture, and Sports, Japan.

References

1. C. Joachim, J.K. Gimzewski, A. Aviram: Nature **408**, 541 (2000)
2. V. Balzani, A. Credi, F.M. Raymo, J.F. Stoddart: Angew. Chem., Int. Ed. Engl. **39**, 3349 (2000)

3. J.-M. Lehn: *Supramolecular Chemistry: Concepts and Perspectives* (VCH, Weinheim, 1995)
4. M. Irie (ed.): *Photochromism: Memories and Switches*, Chem. Rev. **100**, whole of issue 5 (2000)
5. G.H. Brown: *Photochromism* (Wiley Interscience, New York, 1971)
6. H. Dürr, H. Bouas-Laurent: *Photochromism: Molecules and Systems* (Elsevier, Amsterdam, 1990)
7. M. Blank, L.M. Soo, N.H. Wasserman, B.F. Erlanger: Science **214**, 70 (1981)
8. M. Irie, Y. Hirano, S. Hashimoto, K. Hayashi: Macromolecules **14**, 262 (1981)
9. S. Shinkai: Pure Appl. Chem. **59**, 425 (1987)
10. K. Ichimura, Y. Suzuki, T. Seki, A. Hosoki, K. Aoki: Langmuir **4**, 1214 (1988)
11. M. Irie: Adv. Polym. Sci. **94**, 27 (1990)
12. M. Irie: Adv. Polym. Sci. **110**, 49 (1993)
13. T. Ikeda, T. Sasaki, K. Ichimura: Nature **361**, 428 (1993)
14. M. Irie, M. Kato: J. Am. Chem. Soc. **107**, 1024 (1985)
15. Y. Atassi, J.A. Delaire, K. Nakatani: J. Phys. Chem. **99**, 16320 (1995)
16. M. Irie: Chem. Rev. **100**, 1685 (2000)
17. M. Irie, K. Uchida: Bull. Chem. Soc. Jpn. **71**, 985 (1998)
18. S.L. Gilat, S.H. Kawai, J.-M. Lehn: J. Chem. Soc., Chem. Commun. 1439 (1993)
19. S.L. Gilat, S.H. Kawai, J.-M. Lehn: Chem. Eur. J. **1**, 275 (1995)
20. M. Takeshita, M. Irie: J. Org. Chem. **63**, 6643 (1998)
21. T. Kawai, T. Kunitake, M. Irie: Chem. Lett. 905 (1999)
22. O. Kahn: *Molecular Magnetism* (VCH, New York, 1993)
23. A.A. Ovchinnikov: Theor. Chim. Acta **47**, 297 (1978)
24. H. Iwamura, N. Koga: Acc. Chem. Res. **26**, 346 (1993)
25. A. Rajca, J. Wongsriratanakul, S. Rajca: Science **294**, 1503 (2001)
26. T. Matsumoto, T. Ishida, N. Koga, H. Iwamura: J. Am. Chem. Soc. **114**, 9952 (1992)
27. S.J. Jacobs, D.A. Shultz, R. Jain, J. Novak, D.A. Dougherty: J. Am. Chem. Soc. **115**, 1744 (1993)
28. K. Matsuda, T. Yamagata, T. Seta, H. Iwamura, K. Hori: J. Am. Chem. Soc. **119**, 8058 (1997)
29. N.L. Frank, R. Clérac, J.-P. Sutter, N. Daro, O. Kahn, C. Coulon, M.T. Green, S. Golhen, L. Ouahab: J. Am. Chem. Soc. **122**, 2053 (2000).
30. W.T. Borden, E.R. Davidson: J. Am. Chem. Soc. **99**, 4587 (1977)
31. K. Matsuda, H. Iwamura: J. Am. Chem. Soc. **119**, 7412 (1997)
32. K. Matsuda, M. Irie: J. Am. Chem. Soc. **122**, 7195 (2000)
33. K. Matsuda, M. Irie: Chem. Lett. 16 (2000)
34. K. Matsuda, M. Irie: Tetrahedron Lett. **41**, 2577 (2000)
35. B. Bleany, K.D. Bowers: Proc. R. Soc. London A **214**, 451 (1952)
36. K. Matsuda, M. Irie: J. Am. Chem. Soc. **122**, 8309 (2000)
37. K. Matsuda, M. Irie: Chem. Eur. J. **7**, 3466 (2001)
38. R. Brière, R.-M. Dupeyre, H. Lemaire, C. Morat, A. Rassat, P. Rey: Bull. Soc. Chim. France **11**, 3290 (1965)
39. S.H. Glarum, J.H. Marshall: J. Chem. Phys. **47**, 1374 (1967)
40. J.M. Tour: Acc. Chem. Res. **33**, 791 (2000)
41. J.M. Tour: Chem. Rev. **96**, 537 (1996)
42. T. Mitsumori, K. Inoue, N. Koga, H. Iwamura: J. Am. Chem. Soc. **117**, 2467 (1995)
43. K. Matsuda, M. Matsuo, M. Irie: Chem. Lett. 436 (2001)
44. K. Matsuda, M. Matsuo, M. Irie: J. Org. Chem. **66**, 8799 (2001)
45. K. Matsuda, M. Irie: J. Am. Chem. Soc. **123**, 9896 (2001)

3 Single-Molecule Magnets

Keiji Takeda, Kunio Awaga

Summary. Single-molecule magnets (SMMs) are among the best materials for studies of nano-scale magnets. SMMs are located in the boundary region between the discrete atoms and the bulk magnets, and exhibit unusual classical and quantum magnetic properties such as meso-scale magnets. In this chapter, we will review the chemistry and physics of 12-nucleus Mn clusters (Mn_{12}) as a SMM. After explaining the magnetization process and the quantum tunneling or coherence in Mn_{12}, we will describe the magneto-structural correlation in Mn_{12}, referring to the Jahn–Teller isomers of Mn_{12}; a slight difference in structure at the Mn(III) site gives rise to a significant difference in magnetic relaxation. The possible application of SMMs is also described.

3.1 Single-Domain Magnets

Magnetic materials usually exhibit magnetic ordering below a certain critical temperature; this ordering arises from interactions between neighboring spins. In ferromagnets, the magnetic energy favors a parallel spin alignment, while the lowest-energy state is achieved with an antiparallel alignment in antiferromagnets and ferrimagnets. Ferro- and ferrimagnets, which have a spontaneous magnetization, contain magnetic domains. While the atomic (or molecular) magnetizations are all parallel in one domain, the domains have different magnetization orientations so as to reduce the magnetostatic energy. The domains are separated by domain walls in which the magnetization gradually rotates from one direction to another. Domain-wall motion is very important in the technology of magnetic information storage, because it governs the strength of permanent magnets

If an isolated magnetic particle is smaller than ~ 100 Å in diameter, it becomes a single-domain magnet, because it is hard to include domain walls whose widths are ~ 50 Å [1,2]. Recently, the magnetic properties of single-domain magnets have been investigated extensively, making use of the rapid advances in nanostructure fabrication techniques and of highly-sensitive magnetic measurements.

The magnetic dynamics of single-domain particles are simpler than those of bulk magnets. The energy of a single-domain particle depends only on the direction of the magnetization vector. If there is a magnetocrystalline

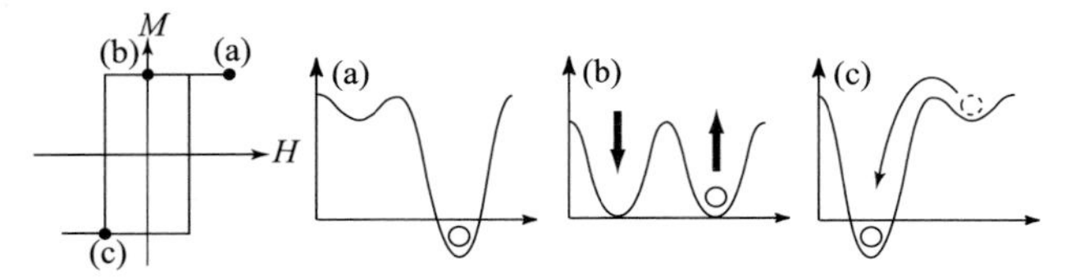

Fig. 3.1. Hysteresis loop in the magnetization curve for single-domain particles. The insets (**a**), (**b**), and (**c**) show the magnetic energy as a function of the magnetization direction at three representative points on the hysteresis loop

anisotropy in this particle, the two stable states, "up" and "down", are separated by a potential barrier, ΔE (see Fig. 3.1). This results in the generation of a hysteresis loop in the magnetization curve, similar to those of bulk ferromagnets. While the hysteresis loops of bulk ferromagnets originate from the motion of domain walls, those of single-domain magnets result from the potential barriers. Figure 3.1 schematically illustrates the field dependence of the double-minimum potential well: (a) the particle stays in the "up" state, exhibiting the saturation magnetization M_s in a large external field applied parallel to the magnetic easy axis; (b) the particle persists in the "up" state because of the barrier, even when the field is reduced to zero; and (c) the magnetization changes its direction from metastable "up" to stable "down" in a field that is opposite to the original direction and is larger than the coercive field H_c.

The relaxation rate $1/\tau$ is governed by the Arrhenius thermal-activation law; the rate is proportional to $\exp(-\Delta E/k_B T)$ at a finite temperature, where k_B is the Boltzmann constant. At high temperatures ($k_B T \geq \Delta E$) the magnetization flips easily because the thermal energy is large enough to overcome the barrier, while the magnetization rotation gradually becomes frozen with a decrease in temperature. We usually define the blocking temperature, T_B, as the point at which $\tau = 100$ s. Hysteresis loops become observable below T_B in conventional magnetic measurements. The value of ΔE is roughly proportional to the product of the magnetic anisotropy and the volume of the particle. There is a dilemma that the smaller volume of the particle, the smaller the values of ΔE and T_B.

Another reason for interest in single-domain magnets is concerned with quantum phenomena [3]. Even below T_B, the magnetization of the particle can rotate by macroscopic quantum tunneling through the barrier. These quantum effects are classified into quantum coherence of magnetization (QCM) [4–8] and quantum tunneling of magnetization (QTM) [9–12]. The former is coherent, while the latter is incoherent. QCM occurs between degenerate states in a symmetric potential well at zero field (Fig. 3.2a); the magnetization rotates coherently between the "up" and "down" states. QTM occurs in an asymmetric potential well. A state on the unstable side makes a transition to an energetically degenerate state on the stable side, going through the barrier, and then relaxes to the ground state (Fig. 3.2b). These

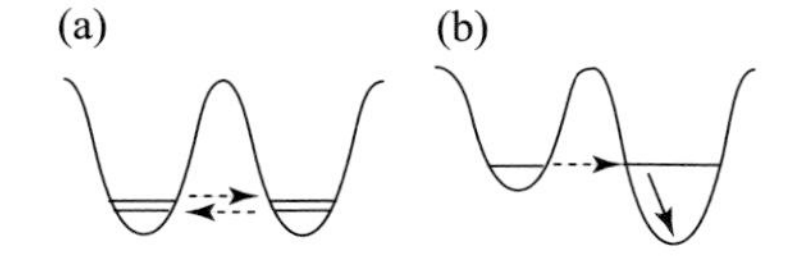

Fig. 3.2. Quantum coherence of magnetization (**a**) and quantum tunneling of magnetization (**b**)

quantum phenomena are being studied extensively, because they are relevant to the performance of nanoscale magnets and to quantum calculations [13].

3.2 Twelve-Nucleus Manganese Clusters

There are physical, chemical, and biological methods to prepare nanoscale magnets. Physical methods, such as lithography, and chemical vapor deposition using the scanning tunneling microscopy techniques [14], have been adopted, especially in the field of applied science. There is an interesting biological method in which an iron-storage protein, ferritin, is utilized as a nanoscale reaction vessel [15–17]. Ferritin is an important metabolic system in animals' spleens, and has a cavity to store iron as ferrihydrite ($5Fe_2O_3 \cdot 9H_2O$). After removal of the content of ferritin samples by dialysis, microparticles of ferrimagnetic magnetite (Fe_3O_4), MnOOH, UO_3, etc., were grown in vitro in the empty protein shell, called apoferritin. The growth reactions were controlled to give particles of a desired size, from a few to a few thousand ferric ions.

In recent years, there has been remarkable development in the chemical preparation of metal-cluster complexes. The nanoparticles obtained from metal-cluster complexes are completely uniform in size and shape, and can form single crystals. Among the compounds prepared, several materials behave as single-domain magnets, showing hysteresis loops in their magnetization curves. These compounds are now called single-molecule magnets (SMMs). The following are examples of SMMs demonstrated so far:

$[Mn^{IV}Mn_3^{III}O_3Cl_4(O_2CCH_3)_3(py)_3]$ [18],
$[Mn_2^{IV}Mn_2^{III}(pdmH)_6(O_2CCH_3)_2(H_2O)_4](ClO_4)_2$ [19],
$[Fe_4^{III}(OCH_3)_6(dpm)_6]$ [20,21],
$[V_4^{III}O_2(O_2CC_2H_5)_7(bipy)_2](ClO_4)$ [22],
$[Mn_4^{II}Mn_3^{III}(teaH)_3(tea)_3](ClO_4)_2 \cdot 3CH_3OH$ [23],
$[Cr\{(CN)Ni(tetren)\}_6](ClO_4)_9$ [24],
$\{[Fe_8^{III}O_2(OH)_{12}(tacn)_6]Br_7 \cdot H_2O\}[Br \cdot 8H_2O]$ [25,26],
$[Fe_{10}^{III}Na_2O_6(OH)_4(O_2CC_6H_5)_{10}(chp)_6(H_2O)_2\{(CH_3)_2CO\}_2]$ [27],
$[Mn_4^{IV}Mn_8^{III}O_{12}(O_2CCH_3)_{16}(H_2O)_4] \cdot 2CH_3CO_2H \cdot 4H_2O$,
$\{[Fe_{17}O_4(OH)_{16}\{N(CH_2CO_2H)_2(CH_2CH_2OH)\}_8(H_2O)_{12}]^+$
$[Fe_{19}O_6(OH)_{14}(N(CH_2CO_2H)_2(CH_2CH_2OH))_{10}(H_2O)_{12}]^+\}$ [28],
$Co_{24}(OH)_{18}(OCH_3)_2Cl_6$(2-methyl-6-hydroxypyridine)$_{22}$ [29],
$[Mn^{IV}Mn_{26}^{IV}Mn_3^{II}O_{24}(OH)_8(O_2CCH_2C(CH_3)_3)_{32}(H_2O)_2(CH_3NO_2)_4]$ [30].

The 12-nucleus Mn clusters are the most extensively studied series among the single-molecule magnets. They have a chemical formula $[Mn_{12}O_{12}(O_2CR)_{16}(H_2O)_4]{\cdot}n$A, where R = CH_3, C_6H_5, etc., and A is a crystal solvent. A great many experimental and theoretical studies have been carried out on these species. In 1980 Lis synthesized $[Mn_{12}O_{12}(O_2CCH_3)_{16}(H_2O)_4]$ $\cdot 2CH_3CO_2H{\cdot}4H_2O$ (abbreviated as Mn_{12}Ac) by a one-pot reaction of $Mn(O_2CCH_3)_2{\cdot}4H_2O$ and $KMnO_4$ in CH_3CO_2H/H_2O [31]. Since then, various Mn_{12} derivatives have been prepared by one-pot reactions similar to that used for Mn_{12}Ac by Lis and also by a ligand substitution method applied to Mn_{12}Ac: stirring Mn_{12}Ac and an excess of a carboxylic acid RCO_2H in CH_2Cl_2 results in an exchange reaction between $CH_3CO_2^-$ and RCO_2^-.

Let us examine the structure of Mn_{12} in detail. Figure 3.3a depicts a schematic view of the core of the Mn_{12} cluster, which consists of four Mn^{4+} and eight Mn^{3+}. Each of the manganese ions is surrounded by six oxygen atoms in an octahedral shape. The central cubane of $[Mn_4^{4+}O_4]^{8+}$ is surrounded by a nonplanar ring that consists of eight Mn^{3+} ions and eight μ_3-O^{2-} ions. This structure has a tetragonal symmetry. The core is additionally coordinated by 16 carboxylate groups and four H_2O ligands. The structures around the Mn^{3+} ions are governed by the Jahn–Teller effect, bringing about elongation of the Mn^{3+} octahedra; one O–Mn–O axis is longer than the other two. These longer O–Mn–O axes are nearly parallel to the molecular axis, which is also the tetragonal axis of the $[Mn_{12}O_{12}]$ core.

The eight Mn^{3+} ions can be classified into two groups, **I** and **II**, as shown in Fig. 3.3a [32]. The group-**I** Mn^{3+} ion is linked to one Mn^{4+} ion and its longer axis is slightly tilted from the molecular axis by $\sim 10°$. In contrast, the group-**II** Mn^{3+} ion is linked to two Mn^{4+} ions and the longer axis makes a significant angle of $\sim 30°$ with the molecular axis. The four

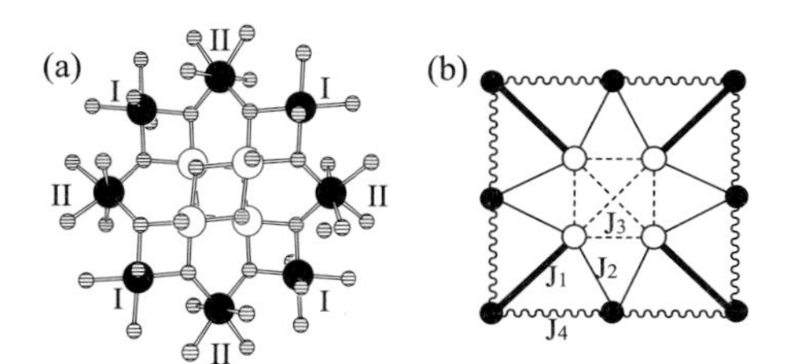

Fig. 3.3. (**a**) Projection of the core of the Mn_{12} cluster along the tetragonal symmetry axis. The central four *open circles* represent Mn^{4+} ions and the outer eight *filled circles* represent Mn^{3+} ions. The *shaded circles* represent the oxygen atoms. The Mn^{3+} ions are classified into groups **I** and **II** (see text). (**b**) The pairwise magnetic exchange interactions in Mn_{12}. The *open* and *filled circles* represent Mn^{4+} and Mn^{3+} ions, respectively. There are at least four types of pairwise exchange interactions; J_1 is between Mn^{4+} and Mn^{3+}, bridged by two μ-oxo ions; J_2 is between Mn^{4+} and Mn^{3+}, bridged by one μ-oxo ion; J_3 is between Mn^{4+} and Mn^{4+}; and J_4 is between Mn^{3+} and Mn^{3+}

H_2O ligands are attached to the group-**II** Mn^{3+} ions in a variety of coordination patterns: four H_2O ligands coordinate to four separate Mn^{3+} ions in $Mn_{12}Ac$ [31], while two H_2O ligands are attached to each of two Mn^{3+} ions in $[Mn_{12}O_{12}(O_2CC_6H_5)_{16}(H_2O)_4]$ (abbreviated as $Mn_{12}Bz$) [33]. Twelve carboxylate ligands bridge the neighboring Mn^{3+} ions on the outside ring and four carboxylate ligands bridge the group-**I** Mn^{3+} ions and the central Mn^{4+} ions.

Cyclic voltammetry measurements on Mn_{12} clusters have revealed two quasi-reversible reductions and one oxidation process: the oxidation–reduction potentials are $E_{\mathrm{re1}} = 0.00$ to $+0.91\,\mathrm{V}$, $E_{\mathrm{re2}} = -0.23$ to $+0.61\,\mathrm{V}$, and $E_{\mathrm{ox1}} = 0.70$ to $1.07\,\mathrm{V}$, respectively. These potentials vary according to the $\mathrm{p}K_{\mathrm{a}}$ of the ligand molecules; the smaller the $\mathrm{p}K_{\mathrm{a}}$value, the easier the reduction of Mn_{12}. Reaction of the neutral Mn_{12} with one or two equivalents of C^+I^- ($C^+ = P(C_6H_5)_4^+$, $N(C_3H_7)_4^+$, etc.) in CH_2Cl_2 or CH_3CN led to one- or two-electron-reduced clusters, as follows:

$$[\mathrm{Mn_{12}O_{12}(O_2CR)_{16}(H_2O)_4}] + n\mathrm{C^+I^-} \rightarrow (\mathrm{C^+})_n[\mathrm{Mn_{12}O_{12}(O_2CR)_{16}(H_2O)}_x]^{n-} + (n/2)\mathrm{I_2}\,, \quad (3.1)$$

where $n = 1$ or 2, and $x = 4$ or 3 [34,35]. X-ray analyses of these salts indicated that one or two of the group-**II** Mn^{3+} ions were reduced to Mn^{2+}, whose sites were trapped. The reduced clusters also behaved as single-molecule magnets, as described in the following sections.

3.3 Thermal Relaxation in Mn_{12}

Figure 3.3b schematically shows the magnetic exchange pathways in $Mn_{12}Ac$. Among the interactions, the strongest is the antiferromagnetic one between Mn^{3+} ($S = 2$) and Mn^{4+} ($S = 3/2$), namely J_1, which leaves up spins on eight Mn^{3+} and down spins on four Mn^{4+}. As a result, Mn_{12} possesses a high-spin ground state of $S = 10$ [33,36,37]. Although the energy gaps between the excited states with different spin multiplicities are not very large, this molecule can be regarded as an $S = 10$ spin species at low temperatures. This cluster also exhibits a strong uniaxial magnetic anisotropy along the tetragonal symmetry axis, caused by the Jahn–Teller distortions at the Mn^{3+} sites [31]. The combination of the high spin nature and the magnetic anisotropy results in a potential barrier between the up and down spin states, as shown in Fig. 3.4 [36]. Rotation of the high-spin magnetic moment of the cluster is gradually frozen below 10 K owing to the intramolecular potential barrier and an imaginary component of the ac susceptibility χ'' arises.

Figure 3.5 shows the temperature dependence of χ'' for $Mn_{12}Ac$ reported in [34]. The plots have a maximum at $\sim 5\,\mathrm{K}$, caused by the freezing effect. By assuming that the relaxation rate $1/\tau$ agreed with the frequency of the ac magnetic field f, i.e. $1/\tau = 2\pi f$, at the temperature of the maximum of χ'', the temperature dependence of $1/\tau$ was obtained. By fitting the data to the

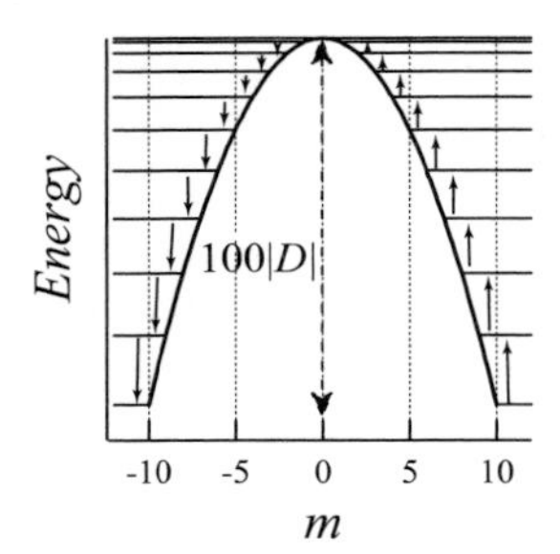

Fig. 3.4. Energy diagram of Mn_{12} ($S = 10$) in zero field, where D is the zero-field splitting. The *solid arrows* indicate allowed transitions from $|+10\rangle$ to $|-10\rangle$

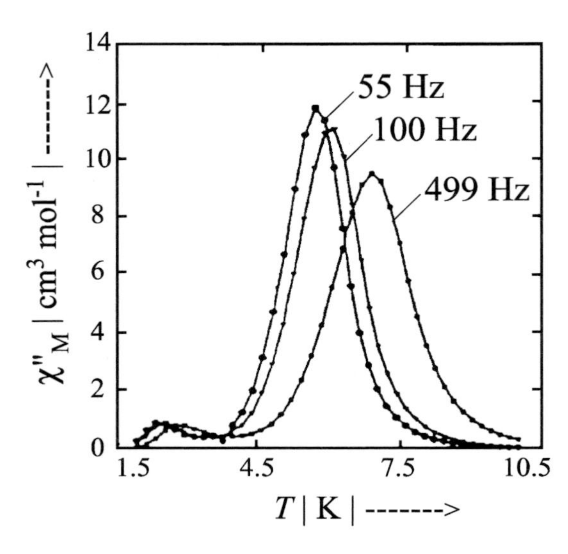

Fig. 3.5. The temperature dependence of the imaginary component χ'' of the ac susceptibility for oriented crystals of $Mn_{12}Ac$ with the ac field parallel to the easy axis basing on [34]

Arrhenius law $\tau = \tau_0 \exp(\Delta E/k_B T)$, the following parameters were obtained for $Mn_{12}Ac$: $\tau_0 = 2.1 \times 10^{-7}$ s and $\Delta E/k_B = 61$ K [37]. On the basis of these parameters, the blocking temperature T_B, where the magnetic relaxation time of the clusters becomes longer than the timescale of dc magnetization measurements ($\tau = 100$ s), was calculated to be 3 K. This value corresponds to the temperature below which a hysteresis loop appears in the magnetization curve. The magnetic interactions between the clusters in the Mn_{12} solids are negligibly weak because of the long distance between neighboring molecules. Therefore the properties described above certainly originate from single molecules. That is why Mn_{12} has been called a single-molecule magnet (SMM). SMMs might supply nanoscale memory devices whose densities are much higher than those of conventional single-domain magnets.

The spin Hamiltonian for $Mn_{12}Ac$ with $S = 10$ and tetragonal symmetry can be written as [38–41]

$$\begin{aligned} \hat{H} = {} & \mu_B \mathbf{B} \cdot \mathbf{g} \cdot \mathbf{S} + D[S_z^2 - (1/3)S(S+1)] \\ & + B_4^0 O_4^0 + B_4^4 O_4^4 + \hat{H}_{\text{dip}} + \hat{H}_{\text{hf}} \ , \end{aligned} \tag{3.2}$$

where S_z is the spin component along the easy axis (here taken as the z axis), D, B_4^0, and B_4^4 are the zero-field splitting parameters, $O_4^0 = 35S_z^4 - 30S(S+1)S_z^2 + 25S_z^2 - 6S(S+1) + 3S^2(S+1)^2$, and $O_4^4 = (1/2)(S_+^4 + S_-^4)$. In the Hamiltonian (3.2), the first term is the Zeeman term, and the second, third, and fourth terms are the anisotropy terms. The terms $\hat{H}_{\mathrm{dip}}$ and $\hat{H}_{\mathrm{hf}}$ represent the dipole interaction with neighboring molecules and the hyperfine interaction with the nuclei, respectively. We can rewrite the Hamiltonian (3.2) as

$$\hat{H} = \alpha S_z^2 - g_z \mu_{\mathrm{B}} B_z S_z + \hat{H}'_z + \hat{H}'_x + \mathrm{const}\ , \tag{3.3}$$

where

$$\hat{H}'_z = \beta S_z^4 - g_z \mu_{\mathrm{B}} B_{dz} S_z \tag{3.4}$$

and

$$\hat{H}'_x = \gamma (S_+^4 + S_-^4) - g_x \mu_{\mathrm{B}} (B_x + B_{dx}) S_x\ . \tag{3.5}$$

Here $\alpha = D - [30S(S+1) - 25]B_4^0$, $\beta = 35B_4^0$, $\gamma = (1/2)B_4^4$, and $B_{\mathrm{d}z}$ and $B_{\mathrm{d}x}$ represent the magnetic fields caused by the contributions of $\hat{H}_{\mathrm{dip}}$ and $\hat{H}_{\mathrm{hf}}$, respectively. $B_{\mathrm{d}z}$ and $B_{\mathrm{d}x}$ vary randomly over space. The third diagonal term commutes with S_z, while the fourth off-diagonal one does not. A high-frequency electron paramagnetic resonance (EPR) measurement, which is one of the most useful methods to determine the parameters in (3.3), gave the values for Mn_{12}Ac as $g_z = 1.93$, $g_x = 1.96$, $\alpha/k_{\mathrm{B}} = -0.56\,\mathrm{K}$, $\beta/k_{\mathrm{B}} = -0.0011\,\mathrm{K}$ and $\gamma/k_{\mathrm{B}} = \pm 2.9 \times 10^{-5}\,\mathrm{K}$ [36].

The unperturbed spin energy is written as $E = \alpha m^2 + \beta m^4 - g_z \mu_{\mathrm{B}} B_z m$, where $m = 0, \pm 1, \pm 2, \cdots, \pm 10$. Since α is negative, $|+10\rangle$ and $|-10\rangle$ are degenerate ground states in zero magnetic field. Now we shall discuss the transition between $|+10\rangle$ and $|-10\rangle$. The first term in (3.3), namely αS_z^2, contains S_+S_+, S_-S_-, S_+S_z, and S_-S_z, so that transitions with $\Delta m = \pm 1$ and ± 2 are allowed. The transition between $|+10\rangle$ and $|-10\rangle$ should occur by stepwise transitions with $\Delta m = \pm 1$ and ± 2, as shown in Fig. 3.4. Since this process is assisted by a spin–phonon interaction, the transition is effectively identical to a thermally activated transition which goes over a potential barrier of $|\alpha S^2 + \beta S^4| = 67\,\mathrm{K}$. This magnetic relaxation can be regarded as an Orbach process, which is well known in the field of paramagnetic resonance [42]. The Hamiltonian (3.3) can explain the potential barrier in Mn_{12}, which has been demonstrated in various magnetic measurements related to the magnetization blocking.

3.4 Jahn–Teller Isomers in Mn_{12}

3.4.1 Magnetic Properties of the Slower and Faster Molecules

Here it is worth noting a complex feature of the magnetization-blocking process. Most of the Mn_{12} clusters exhibit two maxima in the temperature dependence of χ'', at $\sim$ 5 and $\sim$ 2 K, as shown in Fig. 3.5 [33]. The maximum at 5 K is well explained by the potential barrier of 67 K described in Sect. 3.3, but the maximum at 2 K indicates a slower relaxation (SR) process in addition to the faster relaxation (FR) process at 5 K. In contrast to the extensive studies of the SR process, studies of the FR process have been limited, because the higher-temperature maximum is much larger in intensity than the lower-temperature one, i.e. the SR process is dominant in most of the Mn_{12} derivatives characterized so far.

Recently, it was found that the FR process became dominant in (m-MPYNN$^+$)[Mn_{12}Bz]$^-$, where m-MPYNN$^+$ is an organic radical cation [43]. It was demonstrated that the FR process was intrinsic to Mn_{12}, although it had been sometimes regarded as an impurity effect. In addition to this compound, dominance of the FR process was reported in several derivatives [20,44–48]. Among them $Mn_{12}Bz \cdot 2C_6H_5CO_2H$ was useful for investigating the two kinds of relaxation processes; it showed two peaks in χ'' with a strong batch dependence, and successful characterization of the FR process was achieved by analysis of these peaks [44].

Figure 3.6 depicts the temperature dependence of the ac susceptibility for two batches, **A** and **B**, of $Mn_{12}Bz \cdot 2C_6H_5CO_2H$ [48]. These figures show from data on one piece of a crystal from each of Batch **A** and **B**. As shown in Fig. 3.6a, the crystal from Batch **A** showed two peaks in χ'' whose intensities were of the same order of magnitude. Analyses of the frequency dependence of the χ'' maxima with an Arrhenius law $\tau = \tau_0 \exp(\Delta E/k_B T)$ gave the parameters $\tau_0 = 4.7 \times 10^{-9}$ s and $\Delta E/k_B = 66$ K for the SR process, and $\tau_0 = 3.2 \times 10^{-11}$ s and $\Delta E/k_B = 38$ K for the FR process. The blocking temperatures for the SR and FR processes were calculated as $T_B^{SR} = 2.7$ K and $T_B^{FR} = 1.3$ K ($\tau = 100$ s), respectively. Figure 3.6b depicts the results for

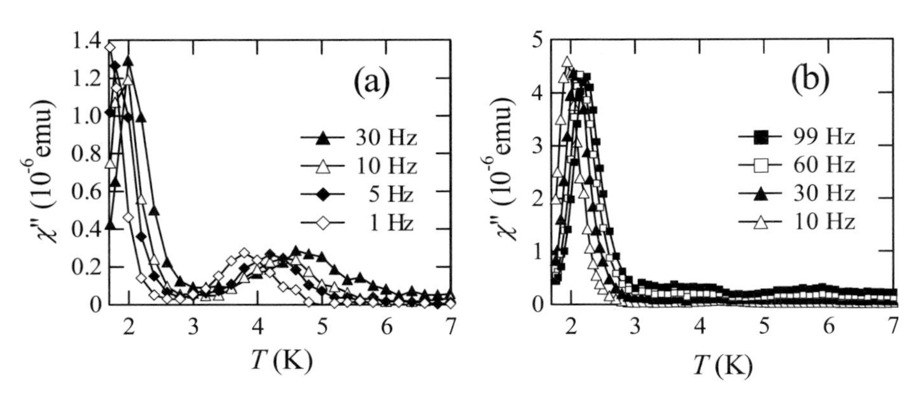

Fig. 3.6. The temperature dependence of the imaginary component χ'' of the ac susceptibility of oriented crystals of $[Mn_{12}O_{12}(O_2CC_6H_5)_{16}(H_2O)_4] \cdot 2C_6H_5CO_2H$ for Batch **A** (**a**) and Batch **B** (**b**) basing on [48]

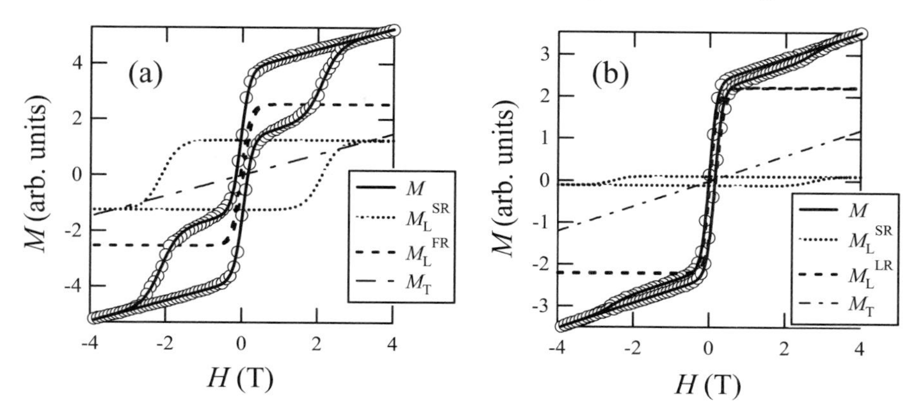

Fig. 3.7. Magnetization curves measured on oriented crystals of Batch **A** at 1.5 K (**a**) and 1.3 K (**b**) basing on [48]

Batch **B**. The plots show a single peak at 2 K; the SR process is negligible in this batch.

Figure 3.7a depicts the magnetization curve for Batch **A** at 1.5 K [48]. The magnetization curve for Batch **A** exhibited hysteresis loops, as did the other Mn_{12} clusters, but their shapes were very unusual. The magnetization curves could be accounted for well assuming the coexistence of two kinds of molecules which were independently governed by the SR and FR processes. As shown in Fig. 3.7a, the magnetization curve is well explained by the sum of three contributions, namely $M = M_{\mathrm{L}}^{\mathrm{SR}} + M_{\mathrm{L}}^{\mathrm{FR}} + M_{\mathrm{T}}$, where $M_{\mathrm{L}}^{\mathrm{SR}}$ and $M_{\mathrm{L}}^{\mathrm{FR}}$ are the longitudinal magnetizations of the SR and FR molecules, respectively, and M_{T} is the sum of the transverse magnetizations of the two. At 1.5 K, between $T_{\mathrm{B}}^{\mathrm{SR}}$ and $T_{\mathrm{B}}^{\mathrm{FR}}$, the rotation of the magnetization is frozen for the SR molecules, so that $M_{\mathrm{L}}^{\mathrm{SR}}$ shows hysteresis. In contrast, the FR molecule does not exhibit hysteresis at this temperature. Figure 3.7b shows the results for Batch **B** at 1.5 K. [48] Since the content of SR molecules was negligible in this batch, the magnetization curve had little hysteresis. It was concluded that the crystals of $Mn_{12}Bz{\cdot}2C_6H_5CO_2H$ had a mosaic structure, at least from the magnetic point of view, and that the ratio of SR and FR molecules was strongly batch-dependent.

3.4.2 Molecular Structure of the Slower and Faster Molecules

To understand the difference between the SR and FR processes, X-ray crystal analyses were carried out on Batch **B**, which included only FR molecules. Figure 3.8a shows a projection of the structure of the $[Mn_{12}O_{12}]$ unit along the molecular axis [48]. A network consisting of Mn^{3+}, Mn^{4+}, and O^{2-} in this structure is quite common for Mn_{12}, but there was an important difference in the environment of Mn^{3+} between this FR molecule and the usual moleucle (namely the SR molecule). Among the eight Mn^{3+} ions, the structures

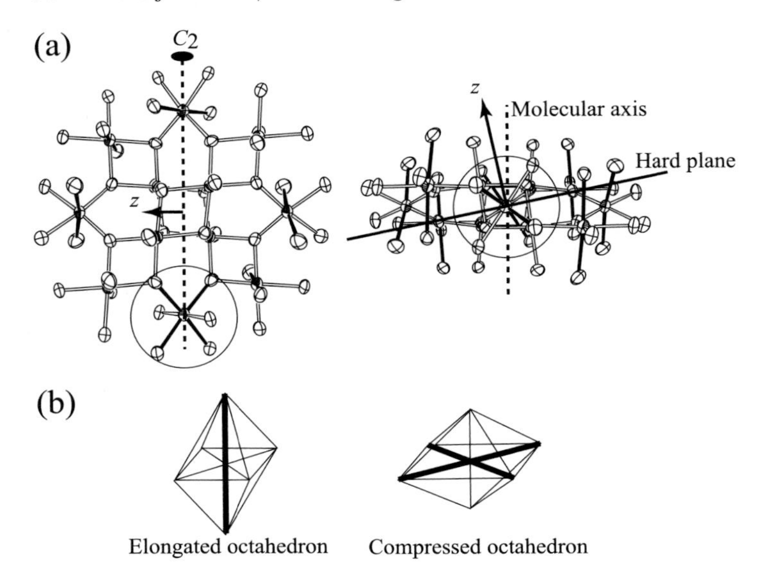

Fig. 3.8. (**a**) Structure of the $[Mn_{12}O_{12}]$ unit (Batch **B**) along the molecular axis (*left*) and the C_2 axis (*right*). (**b**) Schematic views of the structures of the Mn^{3+} site

of seven were of the usual kind, namely elongated octahedra, but that of one (group **II**) was a compressed octahedron in which the short axis was nearly parallel to the molecular axis. The difference between the structures is schematically shown in Fig. 3.8b. Very recently, Sun et al. reported the structures of two Mn_{12} derivatives, in one of which the SR process was dominant and in the other of which the FR process was dominant [45–47]. While the distortions of the Mn^{3+} sites in the former derivative were of the usual kind, i.e. the elongated axes were all nearly parallel to the molecular axis, the latter derivative included an unusual Mn^{3+} site, where the structure was an elongated octahedron but its elongated axis was nearly perpendicular to the molecular axis. The authors called this "Jahn–Teller isomerism" [45–47].

Since the magnetic anisotropy of the whole Mn_{12} molecule is governed by the local distortions at the Mn^{3+} ions, the anisotropy of the FR molecule is significantly different from that of the SR molecule. The angular dependence of the magnetization measured on a Batch **B** crystal suggests that the magnetic easy axis of the FR molecule is tilted by 12° from the molecular axis, as schematically shown in Fig. 3.8a [48].

3.5 Quantum Relaxation in Mn_{12}

3.5.1 Quantum Relaxation of the Slower-Relaxation Molecule

QTM is the most remarkable phenomenon found in Mn_{12} clusters. In this section we describe QTM in the SR molecule. When the magnetic easy axis of $Mn_{12}Ac$, in which the SR molecule is dominant, is oriented parallel to the external field, the magnetization curves below T_B exhibit small steps with an almost constant interval (Fig. 3.9a) [49]. This feature is interpreted as the field-tuned resonant tunneling of magnetization [49–52]. In zero field, the levels $|+m\rangle$ and $|-m\rangle$ in the symmetric double-minimum potential well are degenerate. As the external field increases, the potential well becomes asymmetric. However, the unperturbed energy levels of $|-m+n\rangle$ and $|+m\rangle$ become equal when the external field, B_n, satisfies the following equation:

$$\begin{aligned}\alpha(-m+n)^2 + \beta(-m+n)^4 - g_z\mu_B B_n(-m+n) \\ = \alpha(+m)^2 + \beta(+m)^4 - g_z\mu_B B_n(+m) ,\end{aligned}$$

or

$$\begin{aligned}B_n &= n(1+A_{mn})\alpha/g_z\mu_B , \\ A_{mn} &= (\beta/\alpha)[(m-n)^2 + m^2] . \end{aligned} \quad (3.6)$$

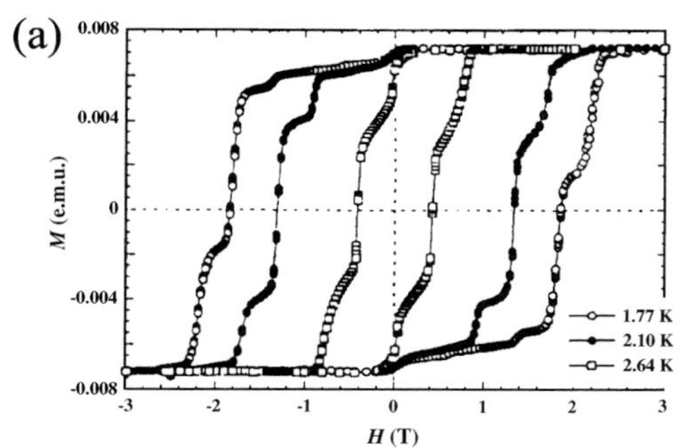

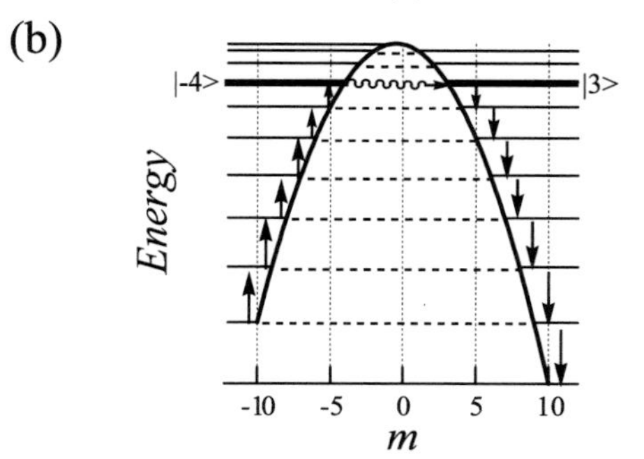

Fig. 3.9. (**a**) Hysteresis loops measured on a single crystal of $Mn_{12}Ac$ in a field parallel to the easy axis at three temperatures. The data was measured at a field sweep rate of 25 Oe/min. (from L. Thomas et al., Nature 383, 145 (1996), reprinted with permission) (**b**) The potential barriers in a field of $H = -\alpha/g\mu_B$. The path shown is that of thermally activated tunneling

As a result, resonant tunneling occurs periodically at specific fields as the field is increased. The observed interval between the resonant values of the field was 0.46 T for $Mn_{12}Ac$. Since $\beta/\alpha \ll 1$, A_{mn} is nearly zero, resulting in the relation $(\alpha/k_B)/g_z \sim 0.30\,K$. This value is consistent with the values $\alpha/k_B = -0.56\,K$ and $g_z = 1.93$, obtained by high-frequency EPR [39]. An important point is that the tunneling phenomena are caused by the term $\hat{H}'_x$ (see (3.5)), which does not commute with S_z [40,53–56]. The existence of this term was demonstrated by high-frequency EPR [39] and by neutron scattering measurements [57]. Although all the unperturbed energy levels $|+m-n\rangle$ and $|-m\rangle$ are degenerate at the resonant field $B_z = B_n$, the diagonal perturbation caused by the random dipolar fields, $g_z\mu_B B_{dz} S_z$, shifts these energies, disturbing the resonance. The term $\gamma(S_+^4 + S_-^4)$ enlarges the splitting due to tunneling, but does not allow a resonance between $|+m-n\rangle$ and $|-m\rangle$ with odd n. The magnetization curve in Fig. 3.9 a shows every possible resonant tunneling with integer n, and it is believed that the transitions of odd n are permitted only by the perturbation by the transverse random dipolar fields, $g_x\mu_B B_{dx} S_x$. Recently, theoretical and experimental studies have indicated that QTM particularly manifests itself between the higher m levels, such as between $|+4\rangle$ and $|-3\rangle$ [51,53–55,40,56,58]. This is called thermally activated tunneling (see Fig. 3.9b).

3.5.2 Quantum Relaxation of the Faster-Relaxation Molecule

QTM in the FR molecule was reported recently. Figure 3.10 presents the magnetic hysteresis curve of the FR molecule in $Mn_{12}Ac$ [59]. Since the concentration of the FR species in the crystal examined was very small, the contribution from the FR molecules was extracted after the demagnetization of the SR molecules. Steps corresponding to resonant tunneling were observed with an

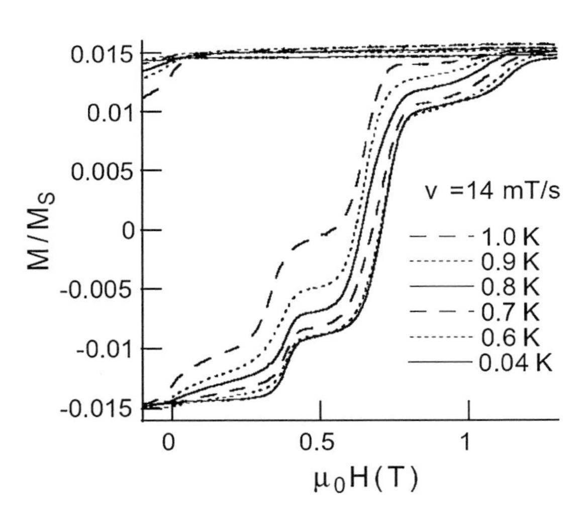

Fig. 3.10. Hysteresis curves of $Mn_{12}Ac$ at various temperatures after demagnetization of the SR molecule (from W. Wernsdorfer et al., Europhys. Lett. 47, 254 (1999), reprinted with permission)

interval of 0.39 T, when the field was parallel to the molecular axis, which is the easy axis of the SR molecule. QTM in the FR molecule was also observed in the Batch **B** crystal of $Mn_{12}Bz \cdot 2C_6H_5CO_2H$, in which the FR molecules were dominant [48]. The magnetization curve of this material was measured in a field making an angle of 37° with the easy axis of the FR molecule. Resonant tunneling of the FR molecule was clearly observed without superposition of the SR contribution. It is expected that the interval would be ~ 0.21 T in a magnetic field parallel to the easy axis. Although there is a difference between the values of the interval for $Mn_{12}Ac$ and $Mn_{12}Bz \cdot 2C_6H_5CO_2H$, these values are both much smaller than that for the SR molecule. As mentioned in Sect. 3.4, the FR molecule has an unusual Mn^{3+} site that probably causes a reduction of the axial anisotropy and an enhancement of the off-axial anisotropy, i.e. $S_x^2 - S_y^2$. The existence of the off-axial anisotropy was demonstrated by EPR measurements on $Mn_{12}Bz \cdot 2C_6H_5CO_2H$ [48]. It is hard to rationalize the quantum effects on the FR molecule at present, on the basis of the structure, but it will be an interesting challenge to do so.

SMMs with off-axial anisotropy (biaxial SMMs) are now a focus of theoretical studies, because Fe_8 clusters with this type of anisotropy exhibit novel QCM in zero longitudinal field; the tunneling splitting width oscillates with an increase in the magnetic field applied along the hard axis, because of a topological quantum interference of the two tunneling paths between the resonant levels [60]. The FR species of Mn_{12}, with its off-axial anisotropy, would be a good material for such studies.

3.6 Quantum Computing with SMMs

It has been demonstrated theoretically that quantum computers are much more efficient than any classical computer for factoring of numbers and database searching [61]. To find a certain number out of N numbers, a classical computer requires $N/2$ queries on average or N at most, while a quantum computer needs fewer than $\sqrt{N}$ queries because of its computational parallelism. A quantum computer consists of "qubits", which contain two quantum states that can interfere. A qubit can represent not only either $|0\rangle$ or $|1\rangle$ but also a superposition of them. Quantum computing is carried out with n coupled qubits, which allow a superposition of 2^n states. For example, carbon-13 labeled chloroform ($CHCl_3$), which contains the nuclear spins of ^1H and ^{13}C, was used as a two-qubit quantum computer by means of NMR techniques [62]. Since each nuclear spin can have an up ($|1\rangle$) or a down ($|0\rangle$) state with respect to the external magnetic field, this can be regarded as a 2-qubit system that contains a superposition of the four states $|00\rangle$, $|10\rangle$, $|01\rangle$ and $|11\rangle$. In this system, only one query was necessary for the desired answer; a query was an application of the appropriate pulsed magnetic field that rotated a nuclear spin, and the read-out was performed by NMR. These measurements were performed in the liquid phase, so that

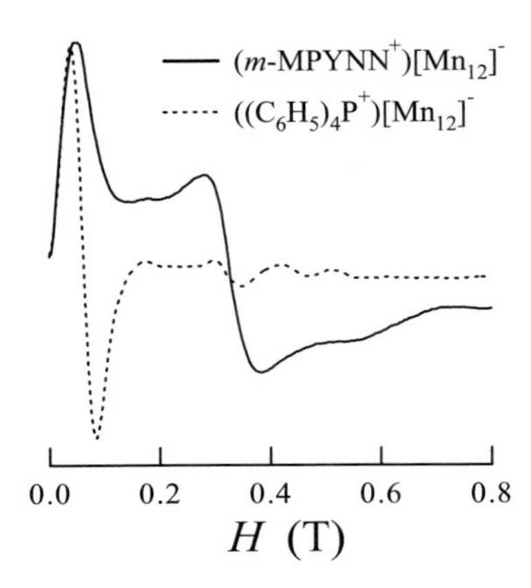

Fig. 3.11. EPR spectra of (m-MPYNN$^+$)[Mn_{12}Bz]$^-$ (*solid curve*) and (($C_6H_5)_4P^+$)[Mn_{12}Bz]$^-$ (*broken curve*) at 12 K

the ensemble nature of the system gave an amplification of the NMR signal without losing the characteristics of a discrete molecule and the coherence of the qubit states.

In crystals of SMMs, the intermolecular exchange interactions are as negligible as those in the $CHCl_3$ liquid. It is therefore possible to create quantum computers using SMMs such as Mn_{12} and Fe_8, in which the two quantum-coherent spin states, $|S\rangle$ and $|-S\rangle$, operate as a qubit. The query and read-out can be carried out by means of pulsed EPR techniques. Coupled n-qubit systems can be realized in supramolecular systems of heterospins, such as (m-MPYNN$^+$)[Mn_{12}Bz]$^-$. Figure 3.11 depicts X-band EPR spectra of (m-MPYNN$^+$)[Mn_{12}Bz]$^-$ and (($C_6H_5)_4P^+$)[Mn_{12}Bz]$^-$ at 12 K, where m-MPYNN$^+$ and $(C_6H_5)_4P^+$ are magnetic and nonmagnetic species, respectively. The absorptions at 0.05 T were assigned to [Mn_{12}Bz]$^-$, while the signal at 0.3 T, which appears only in the m-MPYNN$^+$ salt, was assigned to m-MPYNN$^+$. It is notable that the absorption of [Mn_{12}Bz]$^-$ in the m-MPYNN$^+$ salt is significantly broader than that in the other salt. Since there is a clear overlap between the absorption tails of [Mn_{12}Bz]$^-$ and m-MPYNN$^+$, this broadening can be explained in terms of a cross-relaxation [63–66] caused by a spin–spin interaction. If the organic radical can be replaced with a magnetic species with quantum-coherent spin states, the hybrid material will probably operate as a two-qubit system. Recently it was proposed that the coupled two-qubits could be achieved by physical manipulation of the spin coupling between neighboring SMMs through inductive superconducting loops [67].

3.7 Summary

The thermal and quantum magnetic relaxations of the SMM Mn_{12} have been described in this chapter. Mn_{12} exhibits SR or FR, according to the molecular structure; an unusual distortion at the Mn^{3+} site switches the relaxation from SR to FR. The easy axis of the SR molecule is parallel to the molecular axis, while the easy axis of the FR molecule is tilted by 12°, bringing about a reduction of the axial magnetic anisotropy. The blocking temperature of the FR molecule is less than half of that of the SR molecule. In other words, enhancement of the uniaxial anisotropy causes higher blocking temperatures in SMMs. The highest spin quantum number in the molecule-based magnetic materials that have been studied is $S = {\sim}25$, for $[Mn^{II}\{Mn^{II}(CH_3OH)_3\}_8(CN)_{30}\{Mo^{V}(CN)_3\}_6]$ [68]. If this material has an anisotropy of $D = -0.6\,K$ as Mn_{12} has, the blocking temperature is calculated to be 18 K. The possibility of quantum computing with Mn_{12} was also discussed. An important piece of fundamental rresearch for this purpose would be to elucidate the intermolecular spin–spin interactions of Mn_{12}.

References

1. D.D. Awschalom, D.P. DiVincenzo, J.F. Smyth: Science **258**, 414 (1992)
2. D.D. Awschalom, D.P. DiVincenzo: Phys. Today **48** (3.4), 43 (1995)
3. L. Gunther, B. Barbara (eds.): *Quantum Tunneling of Magnetization: QTM'94* (Kluwer, Dordrecht 1995)
4. A. Grag: Phys. Pev. Lett. **74**, 1458 (1995)
5. D. Garcia-Pablos, N. Garcia, H. De Raedt: Phys. Rev. B **55**, 937 (1995)
6. R. Lu, J.-L. Zhu, X.-B. Wang, L. Chang: Phys. Rev. B **58**, 8542 (1998)
7. S.P. Kou, J.Q. Liang, Y.B. Zhang, F.C. Pu: Phys. Rev. B **59**, 11792 (1999)
8. R. Lu, J.-L. Zhu, J. Wu, X.-B. Wang, L. Chang: Phys. Rev. B **60**, 3435 (1999)
9. G. Scharf, W.F. Wreszinski, J.L. van Hemmen: J. Phys. A: Math. Gen. **20**, 4309 (1987)
10. D.A. Garanin: J. Phys. A: Math. Gen. **24**, L61 (1991)
11. J. Tejada, X. Zhang: J. Magn. Magn. Mater. **140–144**, 1815 (1995)
12. W. Wernsdorfer, E.B. Orozco, K. Hasselbach, A. Benoit, D. Mally, O. Kubo, H. Nakano, B. Barbara: Phys. Rev. Lett. **79**, 4014 (1997)
13. D.P. DiVincenzo: J. Appl. Phys. **81**, 4602 (1997)
14. A.D. Kent, S. von Molnar, S. Gider, D.D. Awschalom: J. Appl. Phys. **76**, 3557 (1994)
15. F.C. Meldrum, B.R. Heywood, S. Mann: Science **257**, 522 (1992)
16. S. Gider, D.D. Awschalom, T. Douglas, S. Mann, M. Chaparala: Science **268**, 77 (1995)
17. J. Tejada, X.X. Zhang, E. del Barco, J.M. Hernandez: Phys. Rev. Lett. **79**, 1754 (1997)
18. S.M.J. Aubin, M.W. Wemple, D. Adams, H.-L. Tsai, G. Christou, D.N. Hendrickson: J. Am. Chem. Soc. **118**, 7446 (1996)
19. J. Yoo, E.K. Brechin, A. Yamaguchi, M. Nakano, J.C. Huffman, A.L. Maniero, L.-C. Brunel, K. Awaga, H. Ishimoto, G. Christou, D.N. Hendrickson: Inorg. Chem. **39**, 3615 (2000)
20. S.M.J. Aubin, N.R. Dilley, L. Pardi, J. Krzystek, M.W. Wemple, L.C. Brunel, M.B. Maple, G. Christou, D.N. Hendrickson: J. Am. Chem. Soc. **120**, 4991 (1998)

21. A.L. Barra, A. Caneschi, A. Cornia, F.F. de Biani, D. Gatteschi, C. Sangregorio, R. Sessoli, L. Sorace: J. Am. Chem. Soc. **121**, 5302 (1999)
22. S.L. Castro, Z. Sun, C.M. Grant, J.C. Bollinger, D.N. Hendrickson, G. Christou: J. Am. Chem. Soc. **120**, 2365 (1998)
23. B. Pilawa, M.T. Kelemen, S. Wanka, A. Geisselmann, A.L. Barra: Europhys. Lett. **43**, 7 (1998)
24. N. Vernier, G. Bellesa, T. Mallah, M. Verdaguer: Phys. Rev. B **56**, 75 (1997)
25. A.-L. Barra, P. Debrunner, D. Gatteschi, C.E. Schulz, R. Sessoli: Europhys. Lett. **35**, 133 (1996)
26. W. Wernsdorfer, R. Sessoli: Science **284**, 133 (1999)
27. C. Benelli, J. Cano, Y. Journaux, R. Sessoli, G.A. Solan, R.E.P. Winpenny: Inorg. Chem. **40**, 188 (2001)
28. A.K. Powell, S.L. Heath, D. Gatteschi, L. Pargi, R. Sessoli, G. Spina, F.D. Giallo, F. Pieralli: J. Am. Chem. Soc. **117**, 2491 (1995)
29. E.K. Brechin, S.G. Harris, A. Harrison, S. Parsons, A.G. Whittaker, R.E.P. Winpenny: J. Chem. Soc., Chem. Commun. 653 (1997)
30. M. Soler, E. Rumberger, K. Folting, D.N. Hendrickson, G. Christou: Polyhedron **20**, 1365 (2001)
31. T. Lis: Acta Cryst. B **36**, 2042 (1980)
32. S.M.J. Aubin, Z. Sun, I.A. Guzei, A.L. Rheingold, G. Christou, D.N. Hendrickson: J. Chem. Soc., Chem. Commun. 2239 (1997)
33. R. Sessoli, H.-L. Tsai, A.R. Schake, S. Wang, J.B. Vincent, K. Folting, D. Gatteschi, G. Christou, D.N. Hendrickson: J. Am. Chem. Soc. **115**, 1804 (1993).
34. H.J. Eppley, H.-L. Tsai, N. de Vries, K. Folting, G. Christou, D.N. Hendrickson: J. Am. Chem. Soc. **117**, 301 (1995)
35. M. Solar, S.K. Chandra, D. Ruiz, J.C. Huffman, D.N. Hendrickson, G. Christou: Polyhedron **20**, 1279 (2001)
36. A. Caneschi, D. Gatteschi, R. Sessoli: J. Am. Chem. Soc. **113**, 5873 (1991)
37. R. Sessoli, D. Gatteschi, A. Caneschi, M.A. Novak: Nature **365**, 141 (1993)
38. P. Politi, A. Rettori, F. Hartmann-Boutron, J. Villain: Phys. Rev. Lett. **75**, 537 (1995)
39. A.L. Barra, D. Gatteschi, R. Sessoli: Phys. Rev. B **56**, 8192 (1997)
40. F. Luis, J. Bartolomé, J.F. Fernández: Phys. Rev. B **57**, 505 (1998)
41. J.F. Fernández, F. Luis, J. Bartolomé: Phys. Rev. Lett. **80**, 5659 (1998)
42. J. Villain, F. Hartman-Boutron, R. Sessoli, A. Rettori: Europhys. Lett. **27**, 159 (1994)
43. K. Takeda, K. Awaga: Phys. Rev. B **56**, 14560 (1997)
44. K. Takeda, K. Awaga, T. Inabe: Phys. Rev. B **57**, 11062 (1998)
45. Z. Sun, D. Ruiz, E. Rumberger, C.D. Incarvito, K. Folting, A.L. Rheingold, G. Christou, D.N. Hendrickson: Inorg. Chem. **37**, 4758 (1998)
46. Z. Sun, D. Ruiz, N.R. Dilley, M. Soler, J. Ribas, K. Folting, M.B. Maple, G. Christou, D.N. Hendrickson: J. Chem. Soc., Chem. Commun. 1973 (1999)
47. S.M.J. Aubin, Z. Sun, H.J. Eppley, E.M. Runberger, I.A. Guzei, K. Folting, P.K. Gantzel, A.L. Rheingold, G. Christou, D.N. Hendrickson: Inorg. Chem. **40**, 2127 (2001)
48. K. Takeda, K. Awaga, T. Inabe, A. Yamaguchi, H. Ishimoto, T. Tomita, H. Mitamura, T. Goto, N. Môri, H. Nojiri: Phys. Rev. B **65**, 4424 (2002)
49. J.R. Friedman, M.P. Sarachik, J. Tejada, R. Ziolo: Phys. Rev. Lett. **76**, 3830 (1996)

50. L. Thomas, F. Lionti, R. Ballou, D. Gatteschi, R. Sessoli, B. Barbara: Nature **383**, 145 (1996)
51. J.M. Hernandez, X.X. Zhang, F. Luis, J. Tejada, J.R. Friedman, M.P. Sarachik, R. Ziolo: Phys. Rev. B **55**, 5858 (1997)
52. J.A.A.J. Perenboom, J.S. Brooks, S. Hill, T. Hathaway, N.S. Dalal: Phys. Rev. B **58**, 330 (1998)
53. D.A. Garanin, E.M. Chudnovsky: Phys. Rev. B **56**, 11102 (1997)
54. H.D. Raedt, S. Miyashita, K. Saito, D. Garcia-Pablos, N. Garcia: Phys. Rev. B **56**, 11761 (1997)
55. E.M. Chudnovsky, D.A. Garanin: Phys. Rev. Lett. **79**, 4469 (1997)
56. A. Fort, A. Rettori, J. Villain, D. Gatteschi, R. Sessoli: Phys. Rev. Lett. **80**, 612 (1998)
57. I. Mirebeau, M. Hennion, H. Casalta, H. Andres, H.U. Gudel, A.V. Irodova, A. Caneschi: Phys. Rev. Lett. **83**, 628 (1999)
58. F. Luis, J. Bartolomé, J.F. Fernández, J. Tejada, J.M. Hernandez, X.X. Zhang, R. Ziolo: Phys. Rev. B **55**, 11448 (1997)
59. W. Wernsdorfer, R. Sessoli, D. Gatteschi: Europhys. Lett. **47**, 254 (1999)
60. W. Wernsdorfer, R. Sessoli: Science **284**, 133 (1999)
61. L.K. Grover: Science **280**, 228 (1998)
62. L.K. Chuang, L.M.K. Vandersypen, X. Zhou, D.W. Leung, S. Lloyd: Nature **393**, 143 (1998)
63. A. Abragam, B. Bleaney: *Electron Paramagnetic Resonance of Transition Ions* (Dover, New York 1986), p. 71
64. J.A. Hodges: Physica B **86–88**, 1143 (1977)
65. C.A. Bates, M. Rezki, A. Vasson, A.-M. Vasson, A. Gavaix: J. Phys. C: Solid State Phys. **14**, 2823 (1981)
66. R. Furrer, C.A. Hutchison, Jr.: Phys. Rev. B **27**, 5270 (1983)
67. J. Tejada: Polyhedron **20**, 1751 (2001)
68. J. Larionova, M. Gross, M. Pilkington, H. Andres, H. Stoeckli-Evans, H.U. Güdel, S. Decurtins: Angew. Chem., Int. Ed. **39**, 1605 (2000)

4 Atomic Resolution of Porphyrins: Single-Molecule Observations of Porphyrinoid Compounds by Scanning Tunneling Microscopy

Ken-ichi Sugiura, Hitoshi Miyasaka, Tomohiko Ishii, Masahiro Yamashita

Summary. This chapter reviews single-molecule observations of porphyrinoid molecules. These are versatile materials that have excellent physical properties associated with their expanded π-electron systems, which are likely to find applications in future nanoscience. The resolution of the observed images is strongly dependent on the substrate used.

4.1 Introduction

In the field of synthetic chemistry, the nuclear magnetic resonance (NMR) technique is indispensable for the characterization of new molecules. Today, no synthetic chemist is able to publish his/her scientific results without NMR data. Furthermore, the recent development of inexpensive, high-performance computers, as well as software having a well-designed graphical user interface, has allowed chemists to report new compounds with reliable structural information provided by single-crystal diffraction studies. Novel instrumental analytical methods, including those two powerful techniques, have created new areas in science.

As described in the preface to this book, the final goal of nanoscience is still unclear. One goal may be the discovery of molecules that mimic large-scale integration (LSI) devices, artificial biological systems, molecule-based machines, and/or advanced materials with enhanced nanoscale quantum physical properties. Whatever the final goal may be, the proposed functions should be associated with isolated single molecules. Hence, observation of the state of individual molecules is essential in this research field. The scanning tunneling microscopy (STM) is one of the most powerful tools for this purpose [1]. In this chapter, we review single-molecule observations of porphyrinoid macrocycles using STM techniques.

4.2 Why Porphyrins?

In this section, we review the significance of porphyrinoid macrocycles in single-molecule science. Why porphyrins?

The porphyrins (**1**, Fig. 4.1) are a class of naturally occurring macrocycles that exibit strong light absorption. Porphyrins play important roles in the field of biology, e.g. in the light-harvesting process of green plants, in the photoinduced charge separation at the photosynthetic reaction center as oxygen carriers, and as redox mediators. Given these versatile functions, these molecules are called "pigments of life". In the field of material science, because of its expanded π-electron system, porphyrins are frequently used as a component molecule for molecule-based conductors [2], magnets [3], and light-emitting devices [4]. It is also expected that porphyrins will be used as component molecules of nanosized advanced materials. Furthermore, the size of these molecules, being more than 10 Å [5], is sufficiently large to enable detection by STM. A number of porphyrins and related phthalocyanines visualized by STM are listed in Table 4.2.

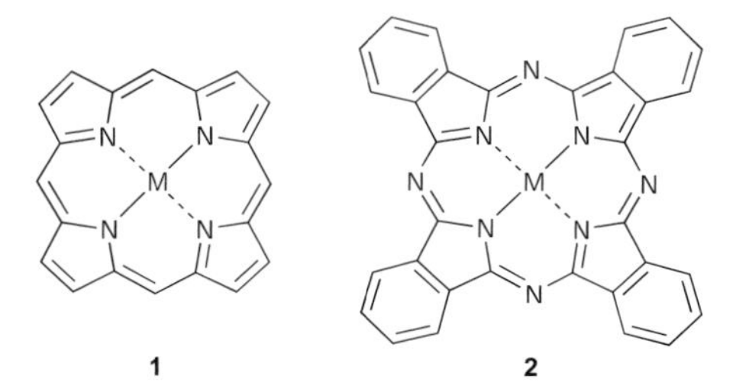

Fig. 4.1. Structures of porphyrin and phthalocyanine

Table 4.1. Lists of the visualized phtahlocyanines and porphyrins by STM

Molecule		Substrate	Reference
Phthalo-cyanines	**2** (M = $Al^{III}Cl$)	graphite	[6]
	2 (M = $Si^{IV}Cl_2$)	H–Si(111)	[7]
	2 (M = $V^{IV}O$)	Au(111)	[8,9]
	2 (M = Fe^{II})	Au(111)	[10]
	2 (M = Co^{II})	Au(111)	[11]
	2 (M = Co^{II}) + **2** (M = Cu^{II})	Au(111)	[11,12]
	2 (M = Ni^{II})	Au(111)	[10]
	2 (M = Ni^{II})	Cu(001)	[13]
	2 (M = Ni^{II})	Si(001)	[13]
	2 (M = Ni^{II})	Si(111)	[14]
	2 (M = Cu^{II})	Cu(100)	[15,16]
	2 (M = Cu^{II})	polycrystalline Ag	[17]

Molecule		Substrate	Reference
	2 (M = Cu^{II})	Ag(111)	[18]
	2 (M = Cu^{II})	Au(111)	[16,19]
	2 (M = Cu^{II})	Si(100)	[13,20–22]
	2 (M = Cu^{II})	Si(111)	[16,20,22,23]
	2 (M = Cu^{II})	H–Si(111)	[24]
	2 (M = Cu^{II})	GaAs	[17]
	2 (M = Cu^{II})	$SrTiO_3$(100)	[25]
	2 (M = Cu^{II})	HOPG	[26–29]
	2 (M = Cu^{II})	ITO	[30]
	2 (M = Cu^{II})	MoS_2	[24]
	2 (M = Cu^{II}) + Perylene	ITO	[30]
	2 (M = Zn^{II})	Si(100)	[31]
	2 (M = Pb^{II})	polycrystalline Au	[32]
	2 (M = Pb^{II})	polycrystalline Si	[33]
	2 (M = Pb^{II})	Si(100)	[34]
	2 (M = Pb^{II})	Si(111)	[34]
	3 (M = Cu^{II})	HOPG	[35]
	3 (M = Zn^{II})	HOPG	[35]
	4 (M = 2H)	Au(111)	[36]
	4 (M = Cu^{II})	Au(111)	[36]
	5 (M = Fe^{II})	MoS_2	[37]
	5 (M = Co^{II})	MoS_2	[37,38]
Porphyrins	**6** (M = Ni)	Au(111)	[39,40]
	6 (M = Co)	Au(111)	[39,40]
	6 (M = Cu)	Au(111)	[39]
	6 (M = Cu)	Cu(111)	[41]
	6 (M = $P^{III}(OCH_2CH_2SH)_2$)	Au–unknown	[42]
	7 (M = 2H)	Au(111)	[43]
	7 (M = Cu)	Cu(100)	[41,44–47]
	7 (M = Cu)	Ag(110)	[45]
	7 (M = Cu)	Au(110)	[45]
	7 (M = Cu)	Cu(111)	[48]
	7 (M = Cu)	Cu(211)	[49]
	7 (M = Cu) + **6** (M = Cu)	Cu(100)	[41]
	8 (M = 2H)	Au(111)	[50]
	9 (M = 2H)	Au(111)	[50]
	10 (M = 2H)	Au(111)	[50]
	11 (M = 2H)	Au(111)	[51]
	12 (M = 2H)	Au(111)	[52–54]
	12 (M = 2H)	Ag(111)	[55]
	12 (M = 2H)	Pt(100)	[56]

Molecule	Substrate	Reference
12 (M = 2H)	S–Au(111)	[57]
13 + adenine	Cu(111)	[58]
14 (M = 2H)	HOPG	[59,60]
14 (M = Fe)	HOPG	[59,60]
14 (M = Zn)	HOPG	[59,60]
15	Cu(111) or Cu(100)	[61]
16	Cu(111) or Cu(100)	[61]
17	Cu(111) or Cu(100)	[62]
18	Cu(100)	[63]
19	Cu(111)	[64]

4.3 Phthalocyanines

4.3.1 Copper Phthalocyanine

Metallophthalocyanines (MPcs) (**2**) are frequently used in STM studies because these molecules have the following characteristics suitable for single-molecule observation: (1) thermodynamic stability, which enables them to exist in the sublimed state; (2) affordability and availability; (3) a sufficiently large size for resolution by STM; (4) perfect chelation with most of the elements in the central core, producing metal–organic hybrid systems; and (5) bulk intrinsic semiconducting properties [65–68].

To the best of our knowledge, the first clear molecular image having atomic resolution was reported for copper phthalocyanine **2** (M = Cu^{II}) by Chiang and coworkers in 1989 [16], other than the images of a few simple molecules such as CO and benzene. MPcs are considered to be the best candidates for obtaining good atomic-resolution images due to the reasons described above [17,28,32]. In the first report of this molecule, a polycrystalline Ag substrate was used, which prevented the resolution of the molecular shape [17]. The molecule was observed as a cone-shaped hill having a diameter of ~ 12 Å, which is consistent with the calculated value for **2** (M = Cu^{II}). Chiang and coworkers replaced the substrate with an atomically clean Cu(100) surface, and this led to a successful result. The observed image shows a cloverleaf shape resembling the top view of **2** (M = Cu^{II}). Detailed analyses in which the current direction is changed, i.e. from the substrate to the STM tip via a molecule or from the tip to the substrate via a molecule, indicate that the visualized image is attributable to the highest occupied molecular orbital (HOMO) of **2** (M = Cu^{II}). Later, other research groups confirmed the reproducibility of this experiment [15].

Stimulated by this landmark publication, several further ideas for single-molecule observations in this system arose.

4.3.2 Effect of Substrate

The selection of the substrate is significant for these STM experiments. As described above, a polycrystalline Ag substrate gives poor molecular images

[17]; however, an atomically clean Ag(111) substrate provides clear images of **2** (M = Cu^{II}) [18]. The main role of the substrate is to bind and/or adsorb the molecule strongly, preventing any drift during the tip scanning. The order of the activity of the substrate is approximately the same as the order of the density of the atoms, i.e. [active] (110) > (100) > (111) [inert] and [active] Ag > Cu > Au [inert]. For example, a Cu(111) surface binds molecules more strongly than a Cu(100) surface.

Along with the metal substrates described above, *active* substrates, i.e. substrates that interact with the molecule to such an extent as to perturb the electronic structure of the molecule, are also used. A molecule on a silicon substrate, one of the most important materials for semiconductor devices and the most fascinating substrate for future nanoscience applying present semiconductor technology [69], was observed in 1994 [22]. An atomically clean silicon substrate is best characterized by unpaired electrons called "dangling bonds" that have a radical character. These electrons are expected to bind molecules strongly. As pointed out by Chiang and coworkers in their first report [16], Si(111) does not provide images of individual molecules. After several attempts, Kawai and coworkers succeeded in obtaining an atomic-resolution image of **2** (M = Cu^{II}) on Si(100) [20]. The observed bulk image of the molecule shows a cloverleaf shape; however, the image also contains dark and bright parts associated with the substrate. The shape of the surface of Si(100) resembles a ridge in a field. A part of a molecule located exactly on a ridge shows a bright image, whereas other moieties located above a valley give dark signals. Kawai and coworkers further investigated another active substrate, $SrTiO_3$(100) [25]. This substrate also binds **2** (M = Cu^{II}) sufficiently to prevent any positioning drifts during the scanning. In contrast, a hydrogen-terminated Si(111) surface, which is considered to be an inert substrate, produces randomly aligned molecules, indicating a decreased activity of the surface [24].

Highly oriented pyrolytic graphite (HOPG) is the most convenient and inexpensive substrate for obtaining an atomically clean surface; this can be done by peeling off the contaminated surface with adhesive tape. Several groups reported isolated molecular images of **2** (M = Cu^{II}); however, no atomic-resolution images were obtained [26–29]. Considering the clear image of **3** on HOPG, it is possible that the orientation of **2** (M = Cu^{II}) on HOPG may not be well matched (see below). Graphite-like MoS_2 seems to be a better substrate for **2** (M = Cu^{II}). Clover-shaped individual molecules similar to those seen on copper are observed using this substrate [24].

4.3.3 Effect of Incorporated Metals

Phthalocyanines, as well as porphyrins, are best characterized by their ability to produce chelate complexes with most of the elements. It is therefore interesting to study the changes of the molecular image when the incorporated metal is replaced with another metals.

In the first report of **2** (M = Cu^{II}), Chiang and coworkers pointed out that the center of this molecule is observed as a "hole", i.e. the tunneling current is too small to be detected at the center position of the molecule [16]. Planar four-coordinated Cu(II) has the $(d_{xy})^2(d_{zx})^2(d_{yz})^2(d_{z^2})^2(d_{x^2-y^2})^1$ configuration. The orbitals that have a z-component perpendicular to the molecular plane, particularly the d_{z^2} orbital, do not affect the visualization. Later, many groups attempted to observe the Cu ion of phthalocyanine, i.e. the d_{z^2} orbital of the Cu(II) ion. To resolve this mysterious phenomenon, Hipps and coworkers measured a series of metallophthalocyanines, M = Cu [11,12], Ni [10], Co [11,12], Fe [10], and V=O [8,9]. For Ni and V=O, the centers of the molecules are dark; by contrast, the Co and Fe complexes exhibit bright lobes attributable to the metal ions. It should be pointed out that the V=O complex has a pot-lid shape; the coordinated divalent oxygen atom is located immediately above the molecule like a knob. Nevertheless, the V=O moiety of the molecule was not visualized. Other nonplanar metallophthalocyanines, namely pot-lid-shaped phthalocyanines such as an Al–Cl complex [7] and a wheel-and-axle-shaped $TiCl_2$ complex [7], were also used in an attempt to observe the metal ion.

In connection with the metal-dependent visualized images, there exist some remaining problems to be solved. One of the most important topics is that of the iron porphyrins and phthalocyanines. Needless to say, the iron porphyrinoid complexes have been widely investigated from the biological viewpoint for more than a century and are regarded as one of the most important groups of compounds in science. Their versatile properties, i.e. adsorption of oxygen and/or nitric oxide, spin crossover phenomenon including the intermediate spin state characteristic of porphyrinoid chelate complexes, and diverse redox behavior will be applicable in nanoscience. As described above, the divalent iron phthalocyanine gives a bright lobe at the center of the molecule [10]. The catalytic and/or enzymatic functions of the iron may be revealed by STM in the future. Further studies are expected on naturally occurring iron porphyrins such as protoporphyrin-IX, **14** [59].

4.3.4 Chemically Modified Phthalocyanines

Some chemically modified phthalocyanine derivatives are reviewed herein. In marked contrast to the complicated UHV–STM experiments that require special treatment, single-molecule observation of a phthalocyanine having eight peripheral octyloxy groups, **3** (M = Cu, Zn) (Fig. 4.2), can be performed under ambient conditions [35]. The molecules can be assembled well on an HOPG substrate by an extremely simple procedure that involves the deposition of one droplet of a 1-phenyloctane solution containing **3** on the HOPG, followed by heating at 60°C for five minutes. Interestingly, replacing 1-phenyloctane with toluene leads to the formation of aggregates on the substrate. Presumably, the structure of the solution, i.e. the state of aggregation of the material in the solution, controls the deposited structure. This simple procedure will be applicable in future nanoscience. The introduction of another substituent, -$OCH_2CH_2OCH_2Ph$, in the same position **4** also produces a monolayer, but the image is not so clear [36].

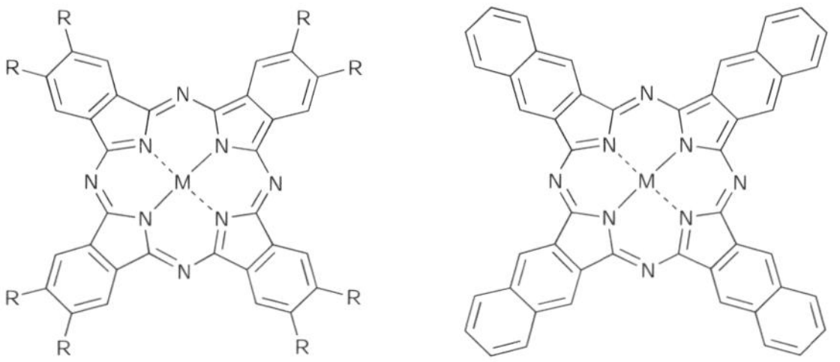

3: R = $O^nC_8H_{17}$

4: R = $OCH_2CH_2OCH_2Ph$

5

Fig. 4.2. Structures of substituted phthalocyanines

Benzene-annulated phthalocyanines, naphthalocyanine **5** (M = Fe and Co), have also been reported; unfortunately, their strong aggregation properties produced a multilayer structure [37,38]. Because the expansion of the π system of phthalocyanine leads to a narrow gap between the HOMO and the lowest unoccupied molecular orbital (LUMO), and their specific D_{4h} symmetry leads to a unique frontier-orbital shape, further detailed investigation of this molecule is expected.

4.4 Porphyrins

Compared with the commercially available phthalocyanines, the sources of porphyrins are limited and expensive. A few simple derivatives, including the corresponding metallocomplexes, are available. However, from the synthetic chemistry viewpoint, there exists thousands of porphyrin synthesis methodologies for preparing the desired molecules. Hence, porphyrins have the potential for fine-scale molecular design that could be applied in nanoscience.

4.4.1 Tetraphenylporphyrins

To the best of our knowledge, the second example of a reliable single-molecule image was obtained for CuTDtBuPP, **7** (Fig. 4.3) [70]. In 1992, we (Sugiura), Dr. Tomohide Takami of Visionarts Research SPINLAB Co., Ltd., Tokyo, Japan, and Professor Yoshiteru Sakata of Osaka University, Osaka, Japan, discussed what types of molecules would give clear molecular images by STM. That was a decade after the discovery of STM [1], and several commercial apparatuses were already on the market. The number of publications describing molecular observations by STM was increasing year by

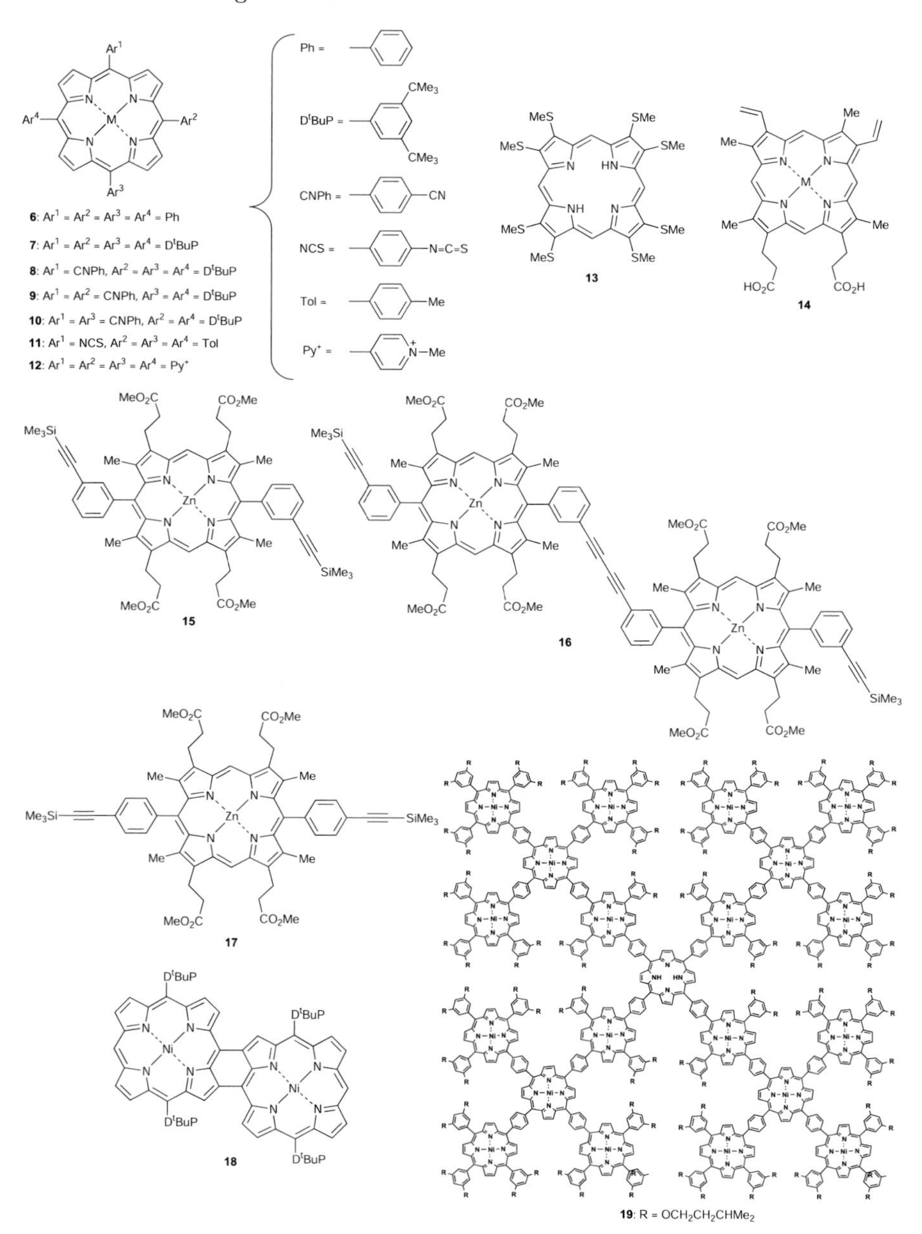

Fig. 4.3. Structures of substituted porphyrins

year. This was a transition period in which physicists handed this analytical tool to chemists. Several reliable reports of atomic resolution were published, especially for self-assembled monolayers (SAMs). These thin layers can be regarded as two-dimensional close-packed arrays of molecules, that give images with periodic stripe patterns. We attempted to obtain a single molecule on a substrate that would show an image recognizable as that of the molecule on sight. To the best of our knowledge, there is only one work that shows a clear image of **2** (M = Cu^{II}) [16]. In this situation, we planned to obtain a clear image of a molecule using STM, which would be easily recognized by everyone.

We reviewed previously reported images in order to propose a new molecular design strategy for a "*visible molecule*", and reconstructed our apparatus into a UHV system to obtain high-resolution images. Although the mechanism of visualization of an atom or a molecule by STM was unclear at that time, as well as at the present time, we proposed the following working hypothesis.

Well-resolved images were reported for SAMs, which are composed of molecules having the alkyl–azobenzene–alkylthiol structure (Fig. 4.4). In the assembled state of these molecules on gold substrates, the molecules

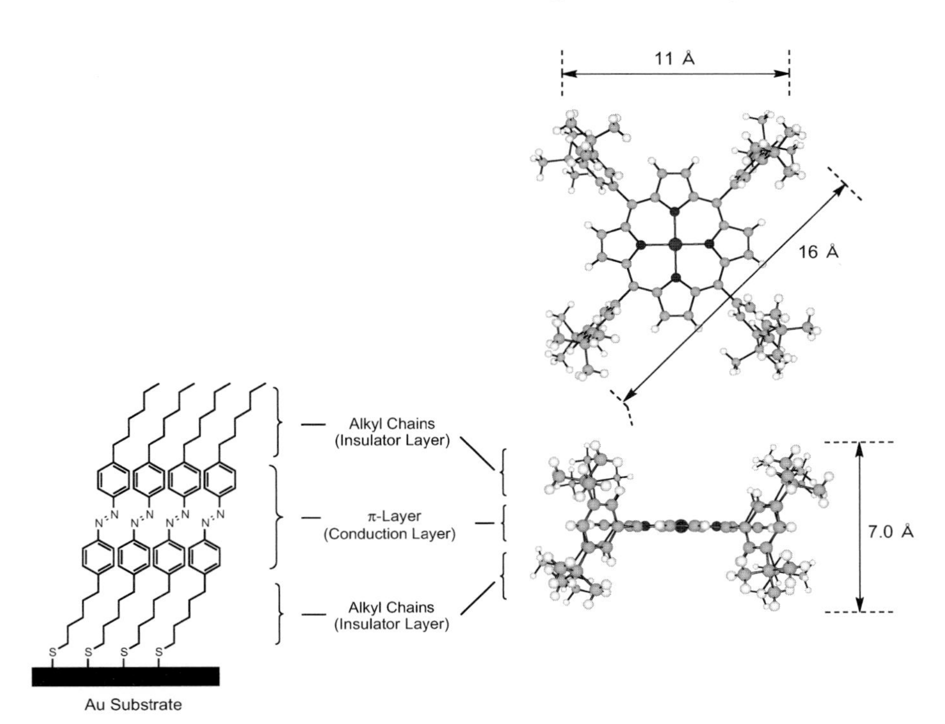

Fig. 4.4. Molecular design for "visualizing molecule", motivated by self-assembled monolayer (SAM)

produce a sandwiched layer structure, i.e. the π-electron layer produced by the azobenzene moiety is sandwiched by two insulating layers of alkyl chain moieties. This structure led us to design an expanded π-electron molecule whose surface is surrounded by alkyl groups, embodying *meso*-tetrakis(3,5-di-*tert*-butylphenyl)porphyrin (TD^{t}BuPP) (**7**). This molecule, reported for the first time by Sasaki and coworkers in 1983 with the aim of elucidating biological redox phenomena [71,72], has four phenyl groups at the *meso*-positions of porphyrin. Usually, the steric repulsion between these phenyl groups and the hydrogen atoms in the β-position produces bond angles of $60 - 80°$, there by preventing π-conjugation between the porphyrin and the phenyl groups [73]. Furthermore, the eight *tert*-butyl groups in the 3- and 5-positions of these phenyl groups, located above and below the sides of the porphyrin mean plane, cover the porphyrin π-system and may act as an insulator. We started our study on this molecule and succeeded in observing the molecular shape with atomic resolution [70]. The molecule was visualized as four bright lobes attributable to the four 3,5-di-*tert*-butylphenyl groups. Stimulated by our successful result, several groups have used this molecule for their single-molecule science programs [43–45,48–50].

Interestingly, the molecular images of **7** measured on an Au(111) surface are strongly affected by temperature. At low temperature, the molecular image consists of a doughnut-shaped open ring surrounded by eight bright lobes. This image is quite different from that measured at room temperature, comprising simply four lobes. As an explanation of these observations, the authors suggested that the dihedral angle between the porphyrin mean plane and the attached phenyl groups decreases to $\sim 20°$ [43]. This value is significantly smaller than that observed for TPP derivatives as revealed by single-crystal diffraction studies, $60 \sim 80°$[73]. The resultant deformation of the porphyrin nucleus makes the internal structure of the porphyrin core visible as a doughnut-shaped open ring. Joachim and coworkers reported a similar observation for the corresponding free base [48]. The porphyrin core moiety of the free base is brighter than that of the corresponding metallocomplex.

Recently, Yokoyama and coworkers succeeded in assembling molecules into patterns based on weak CH–N hydrogen bonds [50]. They replaced some of the 3,5-di-*tert*-butylphenyl groups with 4-cyanophenyl groups (**8-10**). The latter groups act as a supramolecular synthone (SS) to produce aggregates, and the former groups act as "STM-active sites" to visualize the molecule/aggregate. The molecule with one SS produces a triangle (**8**), that with two SSs in *cis*-positions produces a square (**9**), and that with two SSs in *trans*-positions gives a wire (**10**).

As described above, the detailed visualization mechanism of the D^{t}BuP groups is still unclear. However, the image of the molecule with the *tert*-butyl groups removed, *meso*-tetraphenylporphyrin (TPP) **6**, strongly suggests that the D^{t}BuP group acts as an "STM-active" substituent. Hipps and coworkers reported molecular images of **6** and revealed that the molecule gave dim crosses. Applying this "STM-active" property of the D^{t}BuP group, Sugiura and coworkers characterized a novel porphyrin oligomer (**18**) named "porphyracene", as a dimer [63].

4.4.2 Other Simple Porphyrins

The naturally occurring protoporphyrin-IX sodium salt **14** (M = 2H, Fe, and Zn) also produces a highly oriented architecture on HOPG [59]. Furthermore, although the Fe^{II} atom of **2** produces a bright lobe [10], an Fe^{III} atom incorporated into this ligand is observed as a dark hole. In the series based on this ligand, the free base porphyrin gives a highly resolved image of its internal structure.

In 2000, we, Tanaka, and Kawai planned to create an *all*-molecular wiring aimed at a future molecule-based computer [74], i.e. we aimed to produce self-assembled wire shaped molecular alignments between other molecules acting as quantum dots. We chose adenine as a component molecule of the wire and **2** (M = Cu) as a nanodot. Because adenine, which tends to coordinate with metals [25], was used for the wire, the substrate was limited to one of the most inert surfaces, Cu(111). Unfortunately, the deposited **2** (M = Cu) molecules moved as the tip was scanned. The use of sulphur-substituted porphyrin **13** solved this problem, because sulphur is known to interact with copper strongly. After the sublimation of **13** on the substrate, the further deposition of adenine produced an adenine-based nanowire between the nanodots consisting of **13**.

4.4.3 Porphyrin Oligomers

Synthetic studies of oligomeric porphyrins to simulate biological phenomena have been attracting much attention for three decades [75]. Several biological systems show specific porphyrin aggregates, such as a cyclophane-shaped stacked dimer of photosynthetic reaction centers and a wheel-shaped 27-aggregate of a light-harvesting protein [76]. The idea of introducing a photosynthetic mechanism into nanoscience is fascinating. In this context, attempts have been made to characterize some porphyrin oligomers such as **16**, by STM.

The most successful result was reported for the porphyrin 21-oligomer **19** that has a molecular weight of 20, 061 Da ($C_{1244}H_{1350}N_{84}Ni_{20}O_{88}$). The STM image of **19** in its adsorbed state on a Cu(111) surface, prepared by Kawai's pulse injection technique [77], reveals 21 major lobes aligned in a perfect square, with a Mandala pattern where each lobe corresponds to a porphyrinatonickel(II) unit. The size of the square, the Bandanna, is 95 Å along the diagonal direction and 65 Å along the sides, which is consistent with the molecular size of **19** estimated from a molecular-geometry calculation (the calculated values are 93.3 and 65.7 Å). In the molecule, the major lobes are separated by 13 Å in the diagonal direction and by 18 Å along the sides (the calculated values are 12.9 and 18.4 Å). Some of the major lobes are observed as an ensemble of four small lobes attributable to the phenyl groups.

References

1. G. Binning, H. Rohrer, C. Gerber, E. Weibel: Phys. Rev. Lett. **49**, 57 (1982)
2. T.J. Marks: Angew. Chem., Int. Ed. Engl. **29**, 857 (1990)
3. D.K. Rittenberg, K.-i. Sugiura, Y. Sakata, S. Mikami, A.J. Epstein, J.S. Miller: Adv. Mater. **12**, 126 (2000)
4. M.A. Baldo, D.F. O'Brien, Y. You, A. Shoustikov, S. Sibley, M.E. Thompson, S.R. Forrest: Nature **395**, 151 (1998)
5. B.M.L. Chen, A. Tulinsky: J. Am. Chem. Soc. **94**, 4144 (1972)
6. T. Ohmori, H. Masuda, M. Simura, J. Kuroda, T. Okumura: Thin Solid Films **315**, 1 (1998)
7. P. Miao, A.W. Robinson, R.E. Palmer, B.M. Kariuki, K.D.M. Harris: J. Phys. Chem. B **104**, 1285 (2000)
8. K.W. Hipps, D.E. Barlow, U. Mazur: J. Phys. Chem. B **104**, 2444 (2000)
9. D.E. Barlow, K.W. Hipps: J. Phys. Chem. B **104**, 5993 (2000)
10. X. Lu, K.W. Hipps: J. Phys. Chem. B **101**, 5391 (1997)
11. X. Lu, K.W. Hipps, X.D. Wang, U. Mazur: J. Am. Chem. Soc. **118**, 7197 (1996)
12. K.W. Hipps, X. Lu, X.D. Wang, U. Mazur: J. Phys. Chem. **100**, 11207 (1996)
13. G. Dufour, C. Poncey, F. Rochet, H. Roulet, S. Iacobucci, M. Sacchi, F. Yubero, N. Motta, M.N. Piancastelli, A. Sgariata, M. De Crescenzi: J. Electron Spectrosc. Relat. Phenom. **76**, 219 (1995)
14. L. Ottaviano, S. Santucci, S. Di Nardo, L. Lozzi, M. Passacantando, P. Picozzi: J. Vac. Sci. Technol. A **15**, 1014 (1997)
15. P. Sautet, C. Joachim: Surf. Sci. **271**, 387 (1992)
16. P.H. Lippel, R.J. Wilson, M.D. Miller, C. Wöul, S. Chiang: Phys. Rev. Lett. **62**, 171 (1989)
17. J.K. Gimzewski, E. Stoll, R.R. Schlittler: Surf. Sci. **181**, 267 (1987)
18. J.Y. Grand, T. Kunstmann, D. Hoffmann, A. Haas, M. Dietsche, J. Seifritz, R. Möller: Surf. Sci. **366**, 403 (1996)
19. I.I. Smolyaninov: Surf. Sci. **364**, 79 (1996)
20. M. Kanai, T. Kawai, K. Motai, X.D. Wang, T. Hashizume, T. Sakura: Surf. Sci. **329**, L619 (1995)
21. Y. Maeda, T. Matsumoto, M. Kasaya, T. Kawai: Jpn. J. Appl. Phys. Pt. 2 **35**, L405 (1996)
22. F. Rochet, G. Dufour, H. Roulet, N. Motta, A. Sgarlata, M.N. Piancastelli, M. De Crescenzi: Surf. Sci. **319**, 10 (1994)
23. R. Hiesgen, M. Räbisch, H. Böttcher, D. Meissner: Sol. Energy Mater. Sol. Cells **61**, 73 (2000)
24. M. Nakamura, Y. Morita, H. Tokumoto: Appl. Surf. Sci. **113/114**, 316 (1997)
25. H. Tanaka, T. Kawai: Jpn. J. Appl. Phys. Pt. 1 **35**, 3759 (1996)
26. C. Dekker, S.J. Tans, B. Oberndorff, R. Meyer, L.C. Venema: Synth. Met. **84**, 853 (1997)
27. W. Mizutani, Y. Sakakibara, M. Ono, S. Tanishima, K. Ohno, N. Toshima: Jpn. J. Appl. Phys. Pt. 2 **28**, L1460 (1989)

28. W. Mizutani, M. Shigeno, Y. Sakakibara, K. Kajimura, M. Ono, S. Tanishima, K. Ohno, N. Toshima: J. Vac. Sci. Technol. A **8**, 675 (1990)
29. M. Pomerantz, A. Aviram, R.A. McCorkle, L. Li, A.G. Schrott: Science **255**, 1115 (1992)
30. Rudiono, F. Kaneko, M. Takeuchi: Appl. Surf. Sci. **142**, 598 (1999)
31. Y. Maeda, T. Matsumoto, T. Kawai: Surf. Sci. **384**, L896 (1997)
32. C. Hamann, R. Laiho, A. Mrwa: Phys. Status Solidi A **116**, 729 (1989)
33. O. Pester, A. Mrwa, M. Hietschold: Phys. Status Solidi A **131**, 19 (1992)
34. L. Ottaviano, L. Lozzi, S. Santucci, S. Di Nardo, M. Passacantando: Surf. Sci. **392**, 52 (1997)
35. X. Qiu, C. Wang, S. Yin, Q. Zeng, B. Xu, C. Bai: J. Phys. Chem. B **104**, 3570 (2000)
36. P. Smolenyak, R. Peterson, K. Nebesny, M. Toerker, D.F. O'Brien, N.R. Armstrong: J. Am. Chem. Soc. **121**, 8628 (1999)
37. A. Manivannan, L.A. Nagahara, H. Yanagi, A. Fujishima: J. Vac. Sci. Technol. B **12**, 2000 (1994)
38. L.A. Nagahara, A. Manivannan, H. Yanagi, M. Toriida, M. Ashida, Y. Maruyama, K. Hashimoto, A. Fujishima: J. Vac. Sci. Technol. A **11**, 781 (1993)
39. L. Scudiero, D.E. Barlow, K.W. Hipps: J. Phys. Chem. B **104**, 11899 (2000)
40. L. Scudiero, D.E. Barlow, U. Mazur, K.W. Hipps: J. Am. Chem. Soc. **123**, 4073 (2001)
41. J.K. Gimzewski, T.A. Jung, M.T. Cuberes, R.R. Schlittler: Surf. Sci. **386**, 101 (1997)
42. M.S. Boeckl, A.L. Bramblett, K.D. Hauch, T. Sasaki, B.D. Ratner, J.W. Rogers, Jr.: Langmuir **16**, 5644 (2000)
43. T. Yokoyama, S. Yokoyama, T. Kamikado, S. Mashiko: J. Chem. Phys. **115**, 3814 (2001)
44. T.A. Jung, R.R. Schlittler, J.K. Gimzewski, H. Tang, C. Joachim: Science **271**, 181 (1996)
45. T.A. Jung, R.R. Schlittler, J.K. Gimzewski: Nature **386**, 696 (1997)
46. D. Fujita, T. Ohgi, W.L. Deng, H. Nejo, T. Okamoto, S. Yokoyama, K. Kamikado, S. Mashiko: Surf. Sci. **454–456**, 1021 (2000)
47. J.K. Gimzewski, C. Joachim: Science **283**, 1683 (1999)
48. F. Moresco, G. Meyer, K.-H. Rieder, H. Tang, A. Gourdon, C. Joachim: Appl. Phys. Lett. **78**, 306 (2001)
49. F. Moresco, G. Meyer, K.-H. Rieder, H. Tang, A. Gourdon, C. Joachim: Phys. Rev. Lett. **86**, 672 (2001)
50. T. Yokoyama, S. Yokoyama, T. Kamikado, Y. Okuno, S. Mashiko: Nature **413**, 619 (2001)
51. W. Han, S. Li, S.M. Lindsay, D. Gust, T.A. Moore, A.L. Moore: Langmuir **12**, 5742 (1996)
52. M. Kunitake, N. Batina, K. Itaya: Langmuir **11**, 2337 (1995)
53. N. Batina, M. Kunitake, K. Itaya: J. Electroanal. Chem. **405**, 245 (1996)
54. M. Kunitake, U. Akiba, N. Batina, K. Itaya: Langmuir **13**, 1607 (1997)
55. K. Ogaki, N. Batina, M. Kunitake, K. Itaya: J. Phys. Chem. **100**, 7185 (1996)
56. K. Sashikata, T. Sugata, M. Sugimasa, K. Itaya: Langmuir **14**, 2896 (1998)
57. L.-J. Wan, S. Shundo, J. Inukai, K. Itaya: Langmuir **16**, 2164 (2000)
58. M. Furukawa, H. Tanaka, K.-i. Sugiura, Y. Sakata, T. Kawai: Surf. Sci. **445**, L58 (2000)

59. N.J. Tao, G. Cardenas, F. Cunha, Z. Shi: Langmuir **11**, 4445 (1995)
60. B. Duong, R. Arechabaleta, N.J. Tao: J. Electroanal. Chem. **447**, 63 (1998)
61. N. Bampos, C.N. Woodburn, M.E. Welland, J.K.M. Sanders: Angew. Chem., Int. Ed. Engl. **38**, 2780 (1999)
62. P.J. Thomas, N. Berovic, P. Laitenberger, R.E. Palmer, N. Bampos, J.K.M. Sanders: Chem. Phys. Lett. **294**, 229 (1998)
63. K.-i. Sugiura, T. Matsumoto, S. Ohkouchi, Y. Naitoh, T. Kawai, Y. Takai, K. Ushiroda, Y. Sakata: Chem. Commun. 1957 (1999)
64. K.-i. Sugiura, H. Tanaka, T. Matsumoto, T. Kawai, Y. Sakata: Chem. Lett. 1193 (1999)
65. C.C. Leznoff, A.B.P. Lever (eds.): *Phthalocyanines*, Vol. 1 (VCH, New York 1989)
66. C.C. Leznoff, A.B.P. Lever (eds.): *Phthalocyanines*, Vol. 2 (VCH, New York 1992)
67. C.C. Leznoff, A.B.P. Lever (eds.): *Phthalocyanines*, Vol. 3 (VCH, New York 1993)
68. C.C. Leznoff, A.B.P. Lever (eds.): *Phthalocyanines*, Vol. 4 (VCH, New York 1996)
69. G. Ashkenasy, D. Cahen, R. Chohen, A. Shanzer, A. Vilan: Acc. Chem. Res. **35**, 121 (2002)
70. T. Takami, J.K. Gimzewski, R.R. Schlittler, T. Jung, Ch. Gerber, K. Sugiura, Y. Sakata: 50th National Meeting of the Japanese Physical Society, Kanagawa, March 1995, Abstr., 28p-PSB-29.
71. T.K. Miyamoto, S. Tsuzuki, T. Hasegawa, Y. Sasaki: Chem. Lett., 1587 (1983)
72. T.K. Miyamoto, N. Sugita, Y. Matsumoto, Y. Sasaki, M. Konno: Chem. Lett., 1695 (1983)
73. H.L. Anderson: Chem. Commun., 2323 (1999)
74. G.Y. Tseng, J.C. Ellenbogen: Science **294**, 1293 (2001)
75. A.K. Burrell, D.L. Officer, P.G. Plieger, D.C.W. Reid: Chem. Rev. **101**, 2751 (2001)
76. T. Pullerits, V. Sundström: Acc. Chem. Res. **29**, 381 (1996)
77. H. Tanaka, T. Kawai: J. Vac. Sci. Technol., B **15**, 602 (1997)

Part II

Surface Molecular Systems

5 Carboxylates Adsorbed on $TiO_2(110)$

Hiroshi Onishi

Summary. Experimental results on carboxylate ($RCOO^-$) molecules chemisorbed on an atomically flat $TiO_2(110)$ surface are summarized as a prototype of well-defined organo-oxide interfaces. Recent research has led to remarkable advances in preparing carboxylate monolayers and in characterizing those interfaces by thermal desorption spectroscopy, photoelectron spectroscopy, X-ray absorption spectroscopy, vibrational spectroscopy, electron stimulated desorption, and probe microscopy. The principles of carboxylate adsorption, i.e. dissociation at the surface, a bridge structure of the adsorbate, and spontaneous formation of a (2×1)-ordered monolayer at room temperature, are valid for all alkyl-substituted carboxylates ($R = H$, CH_3, C_2H_5, $C(CH_3)_3$, $C\equiv CH$, C_6H_5, and CF_3) ever examined. Selected topographic images of carboxylate monolayers determined by scanning tunneling microscopy and noncontact atomic force microscopy are presented. Different carboxylates ($RCOO^-$ and $R'COO^-$) can be identified molecule-by-molecule by the microscopes when they are coadsorbed.

5.1 Introduction

Increasing attention has been paid to the interfacial science and technology of metal oxides. Current applications of this class of ceramics are based on their intrinsic properties such as their optical transmittance in the visible and near-infrared regions, high refractive index, high dielectric constant, remarkable magnetism, and biocompatibility. These advantages may be utilized in the forthcoming molecule-scale devices for electronic, chemical, and biochemical applications. However, molecule–oxide interfaces have been far less extensively studied than molecule–metal and molecule–silicon interfaces [1]. Examples of molecular monolayer constructed on an atomically flat oxide substrate are especially limited, although organosilanes [2] and long-chain organic acids [3–12] have been used to form self-assembled monolayer on polycrystalline oxides. There were experimental difficulties in preparing and characterizing a stoichiometric surface even using single-crystalline oxide wafers.

Preparation recipes are now established for low-index planes of several compounds (TiO_2, $SrTiO_3$, ZnO, MgO, NiO, α-Al_2O_3, Cr_2O_3, CeO_2, SnO_2, Cu_2O, WO_3, etc.) [13,14]. Thin oxide films epitaxially synthesized by chemical vapor deposition [15] or by molecular-beam epitaxy [16] are also available.

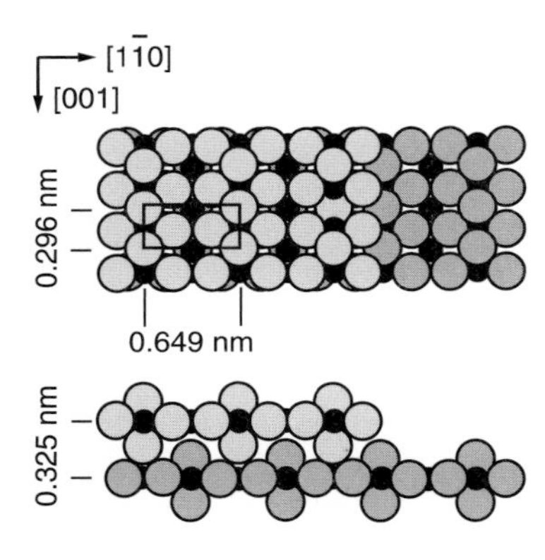

Fig. 5.1. The (1×1) phase of the $TiO_2(110)$ surface. The *black* and *gray circles*, whose radii correspond to the ionic radii, represent Ti and O atoms, respectively. Top and side views of two stacked terraces are shown. One bridge oxygen atom is removed from the first layer for illustration. A (1×1) unit cell is superimposed on the top-view

Single-molecule recognition becomes possible on those surfaces, helped by the vast developments in scanning probe microscopy. The present chapter summarizes experimental results on carboxylate molecules chemisorbed on the $TiO_2(110)$ surface, which serves a prototype of well-defined organo–oxide interfaces.

5.2 Rutile $TiO_2(110)$, the Substrate

Titanium dioxide (TiO_2) is a transition metal oxide used in photofunctional applications such as photocatalysts [17,18], dye-sensitized solar cells [19], and artificial control of hydrophilicity [20]. The constituent Ti and O atoms are fully ionized to the 4+ and 2− oxidation states, leading to a wide band gap of 3.0 eV. High-purity (99.99% or better) single crystals of rutile, one of three polymorphs of this compound, are commercially available. The (110) surface of rutile is the metal oxide surface that has been most extensively studied by surface-sensitive techniques since the 1970s [21]. The (110) surface of the stoichiometric rutile crystal is stable over a wide temperature range up to 1000 K, whereas the (001) [22] and (100) [23] surfaces reconstruct to reduce the surface energy when annealed in a vacuum.

Figure 5.1 illustrates the accepted model for the stoichiometric, unreconstructed (1×1) structure of $TiO_2(110)$ deduced from surface X-ray diffraction [24], medium-energy back-scattered electron diffraction [25], ion scattering [26], and ab initio calculation [27] studies. The topmost layer contains two types of Ti atoms, namely five-fold and six-fold coordinated. The five-fold-coordinated Ti atoms are exposed to the ambient, while the oxygen atoms protruding from the surface plane bridge and cover the six-fold-coordinated Ti atoms. Grooves with exposed Ti atoms and ridges formed by the bridging O atoms make this surface anisotropic. The rectangular surface unit cell of 0.649 nm × 0.296 nm contains one five-fold-coordinated Ti, one six-fold-coordinated Ti, one bridge O, two in-plane O, and one subsurface O atom.

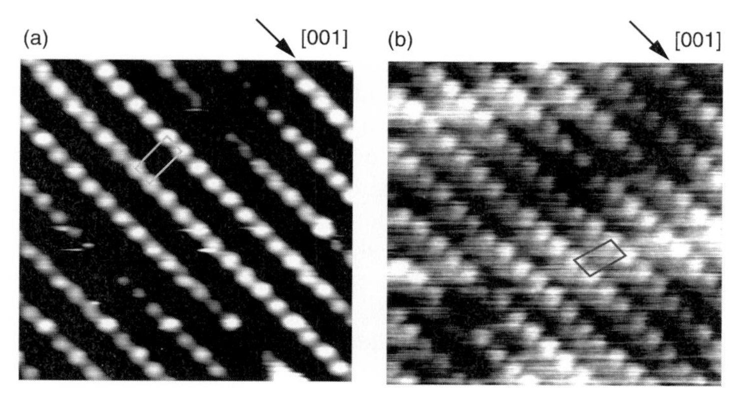

Fig. 5.2. Topographic microscope images of the $TiO_2(110)$ (1×1) surface. (**a**) Constant-current topography observed by STM. Image size $5 \times 5\,nm^2$, sample bias voltage +1.8 V, tunnel current 0.4 nA. (**b**) Constant-frequency-shift topography observed by NC-AFM. Image size $5 \times 5\,nm^2$, frequency shif −68 Hz, oscillation half amplitude 3.4 nm, sample bias voltage +0.9 V. A (1×1) unit cell is superimposed on the images. The slight deviations from a rectanglar shape are due to thermal drift of the surface. The two images were obtained on different samples

A well-ordered surface can be prepared by repeated Ar ion sputtering and vacuum annealing at around 900 K [28]. Vacuum-annealed crystals release oxygen and are reduced to TiO_{2-x}. Titanium atoms in the 3+ state produced in the annealing cause n-type conductivity. Electron-based methods (low-energy electron diffraction, photoelectron spectroscopy, electron energy loss spectroscopy, scanning tunneling microscopy, etc.) are available to characterize the surface. If the oxygen deficiency is small enough, for example a deficiency of the order of 10^{-5} [29], it does not affect the (1×1) long-range order at the surface. More reduced crystals, containing a certain concentration of oxygen vacancies, present several reconstructed surface phases [30,31]. A systematic study of the stoichiometry-dependent phases [32] was performed on heavily reduced crystals.

Efforts at imaging the $TiO_2(110)$ surface by scanning tunneling microscopy (STM) have been made since 1990 [33]. Atom-resolved topography of the (1×1) phase was reported in 1994 [34–36]. It has been established experimentally [28] and theoretically [37] that the unoccupied states localized on the five-fold-coordinated Ti atoms receive electrons from the tip at positive sample bias voltages. The Ti atoms at the bottom of the grooves thus appear as bumps (or a ridge appears as an array of bumps) in the constant-current topographic image of Fig. 5.2a. Such an inversion of the topography does not occur in noncontact atomic force microscopy (NC-AFM). Figure 5.2b shows an NC-AFM topographic image of the (1×1) surface. The bumps ordered parallel to the rectangular unit cells are assigned to the protruding O atoms [38].

Noncontact atomic force microscopy is a rapidly developing scanning probe method that employs a weak attractive force instead of a tunneling current to regulate the tip–surface distance. It offers an opportunity to perform single-molecule sensing on oxides regardless of conductivity. Atom-scale resolution was first achieved on Si [39,40] and InP [41] surfaces in 1995. Details of the noncontact atomic force microscope are described in a recent book [42]. In brief, a cantilever with a tip at the free end is vibrated at its resonance frequency, typically 300 kHz. The peak-to-peak oscillation amplitude of the tip end is several nanometers. A force pulling the tip into the surface causes a reduction of the frequency. If the oscillation amplitude is larger than the tip–surface separation, which is normally the case, the frequency shift is related to a weighted integral of the force over one oscillation cycle [43]. The frequency shift is measured by using a frequency-modulation detection method. The tip–sample distance is regulated to keep the frequency shift constant. A constant-frequency-shift topography is constructed by plotting the tip height as a function of the in-plane coordinates.

5.3 Formate on $TiO_2(110)$

Formate ($HCOO^-$) is an organic adsorbate that is stable on metal oxides and plays an important role as a reaction intermediate in various catalytic reactions. When a $TiO_2(110)$ surface is exposed to formic acid (HCOOH) vapor at room temperature, formates and protons are produced by the dissociation reaction

$$\mathrm{HCOOH(g)} \rightarrow \mathrm{HCOO^-(a)} + \mathrm{H^+(a)}\ , \qquad (5.1)$$

where (g) and (a) represent the gas phase and adsorbate phase, respectively. The formate produced is chemically bound to the surface. The pair of negatively charged oxygen atoms in $HCOO^-$ coordinates two positively charged Ti atoms exposed in a groove. The O–C–O plane of the adsorbate is parallel to the groove axis and the C–H bond is perpendicular to the surface, as illustrated in Fig. 5.3. The O–Ti distance and O–C–O angle were determined from a quantitative analysis of photoelectron diffraction [44] and from a theoretical calculation [45]. The proton released from the acid is thought to combine with a bridge oxygen atom at the surface, $O^{2-}(s)$, to yield a hydroxyl group:

$$\mathrm{H^+(a)} + \mathrm{O^{2-}(s)} \rightarrow \mathrm{OH^-(a)}\ . \qquad (5.2)$$

The dissociative adsorption stops when the five-fold-coordinated Ti atoms are fully occupied. The parent acid molecules physisorbed on the surface vaporize at 180–200 K [46]. A monolayer of chemisorbed formates is hence automatically prepared on a $TiO_2(110)$ surface exposed to the acid vapor at room temperature. The saturated monolayer exhibits a (2×1) long-range order detectable by low-energy electron diffraction. The adsorbate density in

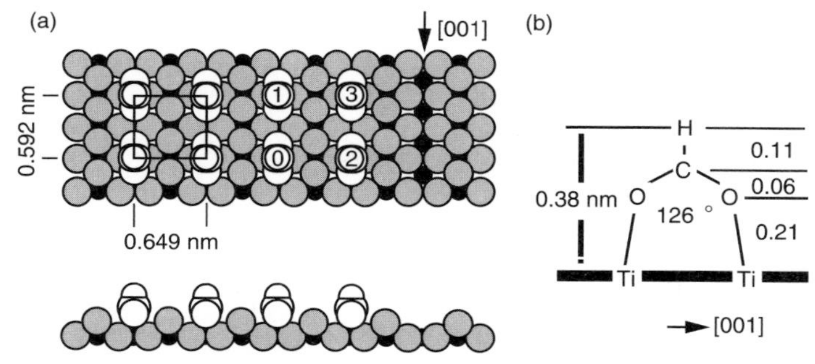

Fig. 5.3. Formates chemisorbed on the $TiO_2(110)$ surface. In (**a**) the white circles, whose radii correspond to the van der Waals radii, represent the formates covering the surface. The right-hand row of the adsorbate is removed for illustration. Top and side views are shown. A (2×1) unit cell is superimposed on the top view. Three formates are numbered in the order of their distance from the formate numbered 0. The proton released in the dissociation reaction is not shown. The atom geometry of the formate is illustrated in (**b**), with atom–atom vertical distances in nm. The O–Ti distance and the O–C–O angle are taken from [44]. Other parameters are assumed to be those of a free HCOOH molecule

the monolayer was estimated as 3×10^{18} molecules m^{-2} on the basis of the X-ray-excited photoelectron intensity of the carbon core level [47]. The model shown in Fig. 5.3 requires a density of 2.61×10^{18} molecules m^{-2}.

When the monolayer-covered surface is heated in a vacuum, 20% of the adsorbate desorbs at 350 K via the reverse reaction of (5.1) [48]. The (2×1) periodicity disappears as the heated surface partially loses the adsorbate. Disproportionation of the hydroxyls to yield water,

$$OH^-(a) + OH^-(a) \rightarrow H_2O(g) + O^{2-}(s) , \tag{5.3}$$

occurs simultaneously. The remaining formates decompose unimolecularly at 570 K:

$$HCOO^-(a) \rightarrow CO(g) + OH^-(a) . \tag{5.4}$$

The products released into the gas phase have been detected in temperature-programmed desorption experiments. The kinetics of the decomposition reaction (5.4) have been analyzed in detail [49]. The stability of the monolayer in air or water is an interesting and important issue that is still to be examined.

An ultraviolet-excited photoelectron spectrum of the chemisorbed formate and physisorbed formic acid has been reported [49]. The work function of the TiO_2 surface decreases by 0.9 eV when a (2×1) formate monolayer covers it [47]. The reduction of the work function is ascribed to the formation of an electric double layer oriented upward, in which the region of highest electron density is located next to the surface. If we neglect the contribution of the hydroxyl produced in reaction (5.2), although this may not be a good assumption, the electric dipole moment of one adsorbed formate can be estimated to be 0.9 debye, oriented towards the vacuum. This estimate of the moment seems reasonable for the formate, the excess electron of which is pulled by the Ti^{4+} atoms and localized at the interface.

Intramolecule vibrations of the formate have been observed by high-resolution electron energy loss spectroscopy (HREELS) [50] and by reflection–absorption infrared spectroscopy (RAIRS) [51]. The symmetric O–C–O stretching mode at a wavenumber of $1365\,cm^{-1}$ and the C–H stretching mode at $2920\,cm^{-1}$ were identified by HREELS. In the RAIR spectrum, the symmetric O–C–O stretching band was observed at $1363\,cm^{-1}$ together with its asymmetric mode at $1566\,cm^{-1}$. In addition, two bands at 1393 and $1535\,cm^{-1}$ were detected by RAIRS and assigned to a minority of the formate with one of its oxygens incorporated into a surface oxygen ridge so as to refill an oxygen vacancy. A X-ray absorption study [52] supports the suggestion of the presence of the minority formate with its O–C–O plane perpendicular to the groove axis, but topography assignable to formate of the assumed geometry has been observed neither by STM nor by NC-AFM.

Formates adsorbed on a TiO_2 surface have been resolved by STM [53] and by NC-AFM [54]. Figure 5.4 presents typical images obtained from the two types of microscope. Adsorbed formates are visible as extra bumps. The STM topography of a fractional (22% of saturation) coverage in Fig. 5.4 a shows that the bumps are located on bright rows (the Ti atom rows) of the substrate. This is consistent with the model of Fig. 5.3. When the surface is saturated with formates, the bumps are aligned on a (2×1) mesh. The anisotropic shape of the individual bumps, which is more noticeable in Fig. 5.4b, is related to the lateral distribution of the molecular orbital receiving electrons from the tip. The lowest unoccupied molecular orbital (LUMO) of a free $HCOO^-$ can be constructed by an antiphase π coupling of the 2p orbitals of the carbon and the two oxygen atoms [55]. The density of the LUMO is enhanced in the direction perpendicular to the O–C–O plane. On the other hand, the van der Waals forces that pull an AFM tip into the molecule [56] are less sensitive to the azimuthal orientation of the formate and create the isotropic topography seen in the NC-AFM image of Fig. 5.4c.

The formate-formate correlation function on the fractionally covered surface of Fig. 5.4a was calculated to provide qualitative information about the intermolecular potential [57]. The occupation probability of a second formate was reduced at the first- and second-nearest-neighbor sites around the first formate. See Fig. 5.3a for the assignment of the sites. Occupation at third-nearest sites was relatively enhanced as a result, and led to a c(4×2) short-range order. The reduced probability to occupy first- and second-nearest-neighbor sites is interpreted as evidence of an electrostatic repulsion between formates, each of which is negatively charged and polarized upward.

On the other hand, the (2×1) long-range order is developed at coverages above 80% of saturation. Close-packed rows of formate must be formed along the grooves on the saturated surface. In the (2×1) structure, two adjacent

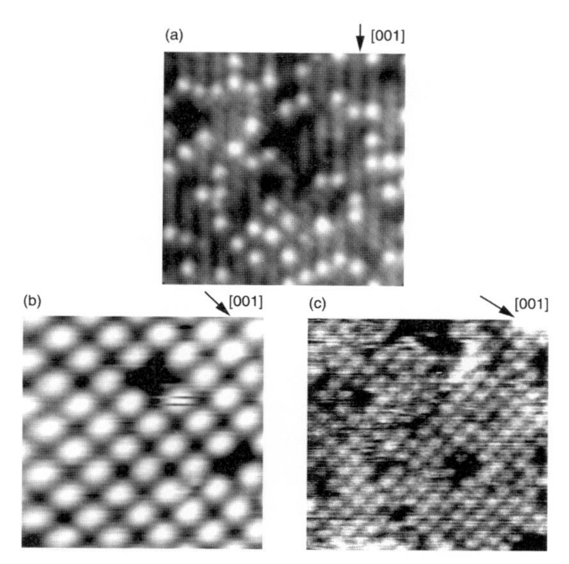

Fig. 5.4. Microscope images of formates on $TiO_2(110)$. (**a**) Constant-current STM topography of isolated formates from ref. [53]. Image size $10 \times 10\,nm^2$, sample bias voltage $+\,1.0$ V, tunnel current 0.2 nA. (**b**) Constant-current STM topography of a (2×1) monolayer from ref. [69]. Image size $5 \times 5\,nm^2$, sample bias voltage $+\,1.3$ V, tunnel current 0.4 nA. (**c**) Constant-frequency-shift NC-AFM topography of a (2×1) monolayer. Image size $10 \times 10\,nm^2$, frequency shift -132 Hz, oscillation half amplitude 3.4 nm, sample bias voltage $+0.4$ V

formate rows are matched in phase. What mediates between two formate rows across an oxygen ridge? The formate–formate repulsion found in a dilute overlayer favors an out-of-phase arrangement to minimize the repulsion and would result in a c(2×2) order instead of the (2×1) order. The proton released in reaction (5.1) is proposed to mediate between the two formate rows. It is thought to combine with a bridge oxygen atom to yield a hydroxyl group between two formates. The Coulomb attraction between the hydroxyl group and the formates stabilizes formate–hydroxyl–formate chains that are as short as possible in the direction perpendicular to the groove.

The electrostatic repulsion of neighboring formates drives them to migrate along the grooves. Figure 5.5 shows time-lapse STM observations of molecules migrating at room temperature [58]. A $14 \times 14\,nm^2$ void (uncovered surface) was created in a (2×1) monolayer at $t = 0$ min by applying an excess bias voltage to the surface. The strengthened electric field induces decomposition of formates beneath the tip nonthermally [29]. The remaining formates diffused into the void and completely refilled it at $t = 63$ min. The square void shrank, with its boundary always being clear and straight. The borders perpendicular to the groove shifted at a velocity of $0.15\,nm\,min^{-1}$, whereas those parallel to the groove did not move at all. The suppression of the shift

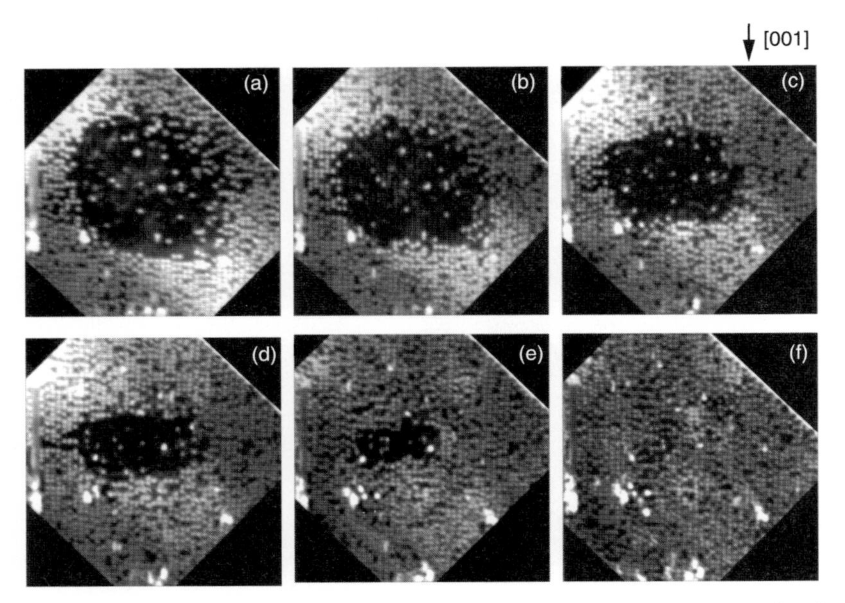

Fig. 5.5. Time-lapse STM images of formates migrating over the $TiO_2(110)$ surface based on refs. [58,59]. Images (**a**)–(**f**) were obtained at t = 10, 15, 26, 35, 50, and 63 min. A void of size $14 \times 14\,nm^2$ was created at $t = 0$. Image size $30 \times 30\,nm^2$, sample bias voltage +1.0 V, tunnel current 0.3 nA

of the parallel borders indicates a significant barrier against movement from one groove to another. A small number of formates had been left in the void of Fig. 5.5a. Careful examination of the sequential images showed that the isolated formates stayed at their original sites until they merged into the migrating monolayer [59]. This demonstrates that the repulsion of neighbors drives the migration.

5.4 Other Carboxylates

The principles of formate adsorption, i.e. dissociation at the surface, the bridge structure of the adsorbate, and spontaneous formation of a (2×1)-ordered monolayer, are valid for all alkyl-substituted carboxylates ($RCOO^-$) ever examined on $TiO_2(110)$. Figure 5.6 illustrates some of these carboxylates. An electron-stimulated desorption ion angular distribution (ESDIAD) study [60] found protons emitted via the rupture of the C–H bonds of acetate ($R = CH_3$). The presence of a C–C bond perpendicular to the surface was concluded from the interpretation of the observed distribution. A X-ray absorption study [52] revealed that the O–C–O plane of acetate or of propanate ($R = CH_2CH_3$) is oriented 0–20 degrees away from the surface normal. Microscope topographic images analogous to that of formate were observed for

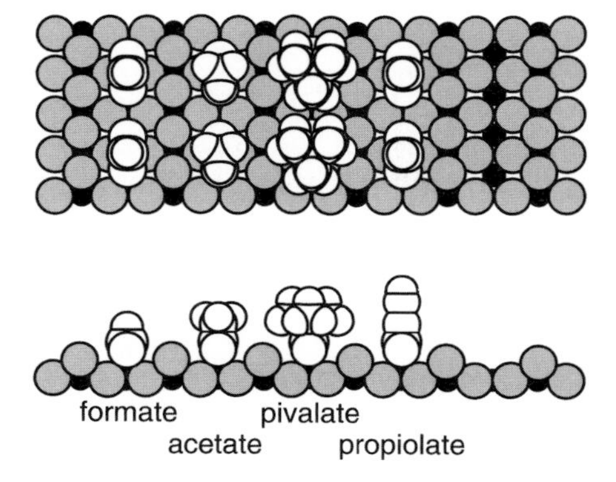

Fig. 5.6. Carboxylates ($RCOO^-$) chemisorbed on the $TiO_2(110)$ surface. The *white circles*, whose radii correspond to the van der Waals radii, represent formate (R = H), acetate (R = CH_3), pivalate (R = $C(CH_3)_3$), and propiolate (R = C≡CH). Top and side views are shown. The proton released in the dissociative adsorption is not shown

acetate [61], pivalate (R = $C(CH_3)_3$) [62], propiolate (R = C≡CH) [62], and trifluoroacetate (R = CF_3) [63]. STM and NC-AFM topographic images of the (2×1) acetate monolayer are presented in Fig. 5.7.

We can thus expect that the ionic O–Ti bonds between the adsorbate and substrate are sufficiently stable for those carboxylates to maintain the bridge structure of the adsorbate. The Ti atoms exposed in the grooves pin the adsorbates and regulate their lateral spacings to 0.65 and 0.59 nm. The interaction between the alkyl groups separated by such a large distance plays a minor role in determining the adsorbate structure. Benzoate (R = C_6H_5) provides an exception. The STM topography of a benzoate monolayer on $TiO_2(110)$ was interpreted as showing the pairing of two aromatic rings across on oxygen ridge [64]. What happens for carboxylates with longer alkyl chains or more aromatic rings is an interesting question to be answered by a systematic study.

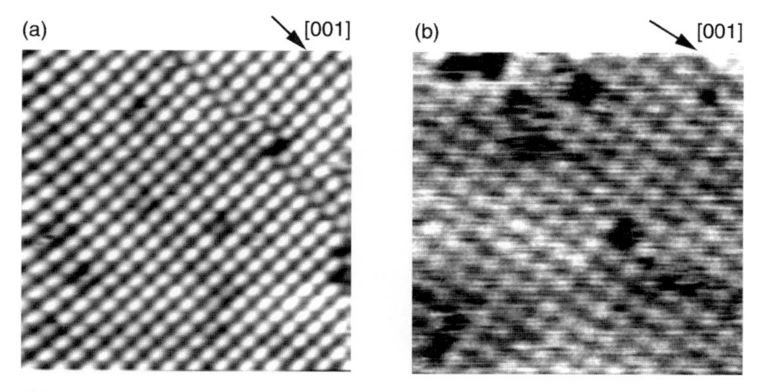

Fig. 5.7. Microscope images of an acetate monolayer on $TiO_2(110)$. (**a**) Constant-current STM topography. Image size $12 \times 12\,nm^2$, sample bias voltage $+1.6$ V, tunnel current 0.4 nA. (**b**) Constant-frequency-shift NC-AFM topography from ref. [63]. Image size $10 \times 10\,nm^2$, frequency shift -88 Hz, oscillation half amplitude 3.4 nm, sample bias voltage -0.2 V

5.5 Carboxylate Monolayers of Mixed Composition

Molecule-scale devices will be composed of different kinds of molecules, which may be electroconductive, nonconductive, photodriven, photoemitting, responsive to particular compounds, etc. These molecules will need be placed at the desired sites on a substrate to make circuits. This section shows how carboxylates with different terminal groups are coadsorbed on $TiO_2(110)$, and how microscopes can resolve those molecules.

A mixed monolayer containing formate and acetate is easily prepared by exposing a formate-covered surface to acetic acid vapor at room temperature. The preadsorbed formate is exchanged with acetic acid arriving from the gas phase [65]. A general reaction formula is

$$RCOO^-(a) + R'COOH(g) \rightarrow R'COO^-(a) + RCOOH(g) . \tag{5.5}$$

The number of acetates is controlled by the exposure time to acetic acid, while the total (acetate + formate) coverage is maintained at the saturation value. Figure 5.8 presents STM [66] and NC-AFM [67,68] topographic images of mixed monolayers at different ratios of the two carboxylates. Individual carboxylates can be clearly identified, like pieces on a checkerboard. The bright and dark adsorbates are assigned to acetate and formate, respectively, because the number of bright molecules increased with exposure to acetic acid vapor. The (2×1) order is maintained in the mixed monolayers independently of the acetate/formate ratio. This is evidence that the two species are adsorbed on the surface in the same way, in the bridge geometry, despite the different terminal groups. The carboxylates in formate–pivalate [69], formate–propiolate [70], and acetate–trifluoroacetate [63] monolayers prepared similarly were identified molecule-by-molecule by NC-AFM topography. The

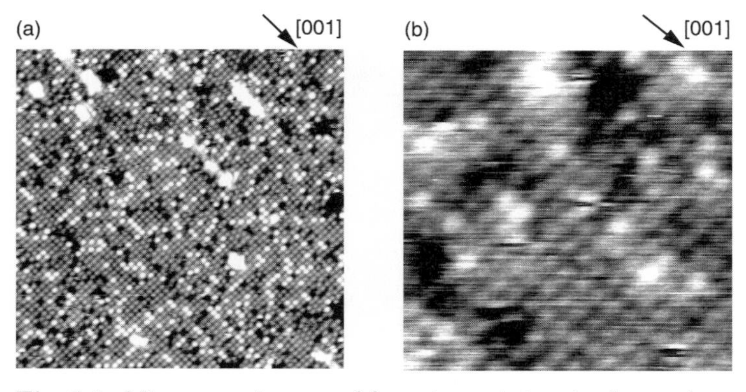

Fig. 5.8. Microscope images of formate–acetate mixed monolayers on $TiO_2(110)$. (**a**) Constant-current STM topography from ref. [66]. Image size $28\times28\,nm^2$, sample bias voltage +1.7 V, tunnel current 0.4 nA. (**b**) Constant-frequency-shift NC-AFM topography of a (2×1) monolayer from ref. [67]. Image size $10\times10\,nm^2$, frequency shift −106 Hz, oscillation half amplitude 3.6 nm, sample bias voltage +0.3 V

molecule-dependent NC-AFM topography is related to the physical topography of R = H, CH_3, $C(CH_3)_3$, and C≡CH quantitatively [56]. The strongly polarized CF_3 group in trifluoroacetate perturbs the tip–molecule force via electrostatic coupling [63]. This demonstrates the ability of NC-AFM to probe the electrostatic properties of an organo-oxide interface on a single-molecule scale.

The acetate–acetate correlation function was calculated for acetate–formate mixed monolayers. The function obtained exhibited some deviation from a random distribution [66]. The interaction between the CH_3 groups of two acetates perturbs the lateral distribution of acetates embedded in a formate monolayer. This is a quite interesting phenomenon, which could be extended to the self-assembly of a desired molecule embedded in other molecules.

5.6 Prospects

The surface science of well-defined, single-crystalline metal oxides was initiated in the United States three decades ago. Current researchers in the US, European countries, and Japan have made remarkable advances in the preparation and characterization of organo-oxide interfaces, a limited fraction of which are described in the present account. We now have several issues to be examined for future exploitation of molecule-scale devices on metal oxide substrates.

There is a clear need to study organo-oxide interfaces with a higher level of complexity. Most of the previous studies have been aimed at interpreting catalytic reactions of simple molecules on oxides. The following molecules have been studied on $TiO_2(110)$: Langmuir–Blodgett films of stearic acid [71], bi-isonicotinic acid [72], pyridine [73–75], glycine [76], an organosilane ($CF_3CH_2CH_2Si(OCH_3)_3$) [77], cromyl chloride (CrO_2Cl_2) [78], metal carbonyl complexes of $Rh(CO)_2$ [79,80], $Rh(acac)(CO)_2$ [81], and $Ru_3(CO)_{12}$ [82]. Custom adsorbates synthesized for specific applications will need to be added to this list.

Modifying oxide substrates is another challenge. Surface singularities such as oxygen vacancies [37], single-height steps [83], and deposited hydrogen atoms [84] can be nuclei for anchoring desired molecules. Scanning tunneling spectroscopy is a powerful tool for mapping site-specific electronic states on oxide surfaces [85]. Sulfurization, as reported for $TiO_2(110)$ [86], may be extended to passivation in reactive atmospheres. Deposited metal clusters have also been extensively studied in relation to catalysis [87–89]. They can potentially be used as electrodes for wiring. In addition, transition metal atoms in oxides have the potential for spin engineering. A transparent ferromagnetic material has been combinatorially synthesized by doping Co into anatase TiO_2 film [90]. Various forms of coupling of electrons, photons, phonons, spins, and chemical substances are possible in tailored organo-oxide interfaces.

Vibrational spectroscopy is of vital importance in characterizing molecules or supramolecules. The experimental methods currently employed for metal oxide substrates have limitations, however. In HREELS, multiple phonon losses in the substrate cause strong peaks, which need to be removed from the spectrum by Fourier deconvolution [91]. Identification of weak but important modes is difficult. The problem of insufficient sensitivity is more serious in the case of RAIRS performed on dielectric oxide substrates, where the signal enhancement available on metals is absent. Sum-frequency (SF) vibrational spectroscopy [92] provides an alternative approach. The latest technical developments make the author optimistic about future application of this method to organo-oxide interfaces. The multiplex detection of broad-bandwidth SF light dramatically increases the signal-to-noise ratio [93], and the use of wavelength-tunable visible light to generate SF signal [94] avoids possible interfereance from the photoluminescence of oxide substrates.

Acknowledgments

The STM images of Fig. 5.4a and Fig. 5.5 were obtained by the author at the University of Tokyo under the supervision of Prof. Y. Iwasawa. Dr. H. Uetsuka and Dr. A. Sasahara of the Kanagawa Academy of Science and Technology obtained the other microscope images shown in the present account. Profitable discussions with Dr. H. Tada and Dr. S. Tanaka of the Molecular Science Institute are acknowledged.

References

1. F. Schreiber: Prog. Surf. Sci. **65**, 151 (2000)
2. J. Sagiv: J. Am. Chem. Soc. **102**, 92 (1980)
3. D.L. Allara, R.G. Nuzzo: Langmuir **1**, 45 (1985)
4. P.E. Laibinis, J.J. Hickman, M.S. Wrighton, G.M. Whitesides: Science **245**, 845 (1989)
5. Y.T. Tao: J. Am. Chem. Soc. **115**, 4350 (1993)
6. E. Smith, M.D. Porter: J. Phys. Chem. **97**, 8032 (1993)
7. A.H.M. Songtag, M.C. Raas: J. Chem. Phys. **91**, 4926 (1989)
8. Y.G. Aronoff, B. Chen, G. Lu, C. Seto, J. Schwartz, S.T. Bernasek: J. Am. Chem. Soc. **119**, 260 (1997)
9. J.G. Van Alsten: Langmuir **15**, 7605 (1999)
10. D. Brovelli, G. Hähner, L. Ruiz, R. Hofer, G. Kraus, A. Waldner, J. Schlösser, P. Oroszlan, M. Ehrat, N.D. Spencer: Langmuir **15**, 4324 (1999)
11. J.T. Woodward, A. Ulman, D.K. Schwartz: Langmuir **12**, 3626 (1996)
12. W. Gao, L. Dickinson, C. Grozinger, F.G. Morin, L. Reven: Langmuir **13**, 115 (1997)
13. V.E. Henrich, P.A. Cox: *The Surface Science of Metal Oxides* (Cambridge University Press, Cambridge 1994)

14. D.A. King, D.P. Woodruff (eds.): *The Chemical Physics of Solid Surfaces*, Vol. 9 (Elsevier, Amsterdam 2001)
15. R. Franchy: Surf. Sci. Rep. **38**, 195 (2000)
16. S.A. Chambers: Surf. Sci. Rep. **39**, 105 (2000)
17. A. Fujishima, K. Honda: Nature **238**, 37 (1972)
18. M.R. Hoffmann, S.T. Märtin, W. Choi, D.W. Bahmanen: Chem. Rev. **95**, 69 (1995)
19. A. Hagfeldt, M. Grätzel: Chem. Rev. **95**, 49 (1995)
20. R. Wang, N. Sakai, A. Fujishima, T. Watanabe, K. Hashimoto: J. Phys. Chem. B **103**, 2188 (1999)
21. U. Diebold: Surf. Sci. Rep. in preparation
22. L.E. Firment: Surf. Sci. **116**, 205 (1982)
23. Y.W. Chung, W.J. Lo, G.A. Somorjai: Surf. Sci. **64**, 588 (1977)
24. G. Charlton, P.B. Howes, C.L. Nicklin, P. Steadman, J.S.G. Taylor, C.A. Muryn, S.P. Harte, J. Mercer, R. McGrath, D. Norman, T.S. Turner, G. Thornton: Phys. Rev. Lett. **78**, 495 (1997)
25. B.L. Maschhoff, J.M. Pan, T.E. Madey: Surf. Sci. **259**, 190 (1991)
26. B. Hird, R.A. Armstrong: Surf. Sci. **420**, L131 (1999)
27. M. Ramamoorthy, D. Vanderbilt, R.D. King-Smith: Phys. Rev. B **49**, 16721 (1994)
28. H. Onishi, K. Fukui, Y. Iwasawa: Bull. Chem. Soc. Jpn. **68**, 2447 (1995)
29. H. Onishi, Y. Iwasawa: Jpn. J. Appl. Phys. **33**, L1338 (1994)
30. H. Onishi, Y. Iwasawa: Phys. Rev. Lett. **76**, 791 (1996)
31. R.A. Bennett, P. Stone, N.J. Price, M. Bowker: Phys. Rev. Lett. **82**, 3831 (1999)
32. M. Li, W. Hebenstreit, U. Diebold, A.M. Tyryshkin, M.K. Bowman, G.G. Dunham, M.A. Henderson: J. Phys. Chem. B **104**, 4944 (2000)
33. G.S. Rohrer, V.E. Henrich, D.A. Bonnell: Science **250**, 1239 (1990)
34. M. Sander, T. Engel: Surf. Sci. **313**, L263 (1994)
35. H. Onishi, Y. Iwasawa: Surf. Sci. **313**, L783 (1994)
36. D. Novak, E. Garfunkel, T. Gustafsson: Phys. Rev. B **50**, 5000 (1994)
37. U. Diebold, J.F. Anderson, K.O. Ng, D. Vanderbild: Phys. Rev. Lett. **77**, 1322 (1996)
38. K. Fukui, H. Onishi, Y. Iwasawa: Phys. Rev. Lett. **79**, 4202 (1997)
39. F.J. Giessibl: Science **267**, 68 (1995)
40. S. Kitamura, M. Iwatsuki: Jpn. J. Appl. Phys. **34**, L145 (1995)
41. Y. Sugawara, M. Ohta, H. Ueyama, S. Morita: Science **270**, 1646 (1995)
42. S. Morita, R. Wiesendanger, E. Meyer: *Noncontact Atomic Force Microscopy* (Springer, Berlin, Heidelberg 2002)
43. M. Guggisberg, M. Bammerlin, C. Loppacher, O. Pfeiffer, A. Abdurixit, V. Barwich, R. Bennewitz, A. Baratoff, E. Meyer, H.J. Güntherodt: Phys. Rev. B **61**, 11151 (2000)
44. S. Thevuthasan, G.S. Herman, Y.J. Kim, S.A. Chambers, C.H.F. Peden, Z. Wang, R.X. Ynzunza, E.D. Tober, J. Morais, C.S. Fadley: Surf. Sci. **401**, 261 (1998)
45. P. Käckell, K. Terakura: Surf. Sci. **461**, 191 (2000)
46. M.A. Henderson: J. Phys. Chem. **99**, 15253 (1995)
47. H. Onishi, T. Aruga, C. Egawa, Y. Iwasawa: Surf. Sci. **193**, 33 (1988) [the adsorbate assignment in this paper was later revised]

48. K.S. Kim, M.A. Barteau: J. Catal. **125**, 353 (1990)
49. H. Onishi, T. Aruga, Y. Iwasawa: J. Catal. **146**, 557 (1994)
50. M.A. Henderson: J. Phys. Chem. B **101**, 221 (1997)
51. B.E. Hayden, A. King, M.A. Newton: J. Phys. Chem. B **103**, 203 (1999)
52. A. Gutiérrez-Sosa, P. Martínez-Escolano, H. Raza, R. Lindsay, P.L. Wincott, G. Thornton: Surf. Sci. **471**, 163 (2001)
53. H. Onishi, Y. Iwasawa: Chem. Phys. Lett. **226**, 111 (1994)
54. K. Fukui, H. Onishi, Y. Iwasawa: Chem. Phys. Lett. **280**, 296 (1997)
55. S.D. Peyerimhoff: J. Chem. Phys. **47**, 349 (1967)
56. H. Onishi, A. Sasahara, H. Uetsuka, T. Ishibashi: Appl. Surf. Sci. **188**, 257 (2002)
57. H. Onishi, K. Fukui, Y. Iwasawa: Jpn. J. Appl. Phys. **38**, 3830 (1999)
58. H. Onishi, Y. Iwasawa: Langmuir **10**, 4414 (1994)
59. H. Onishi, K. Fukui, Y. Iwasawa: Colloids Surf. A **109**, 335 (1996)
60. Q. Guo, I. Cocks, E.M. Williams: J. Chem. Phys. **106**, 2924 (1997)
61. H. Onishi, Y. Yamaguchi, K. Fukui, Y. Iwasawa: J. Phys. Chem. **100**, 9582 (1996)
62. A. Sasahara, H. Onishi, unpublished
63. A. Sasahara, H. Uetsuka, H. Onishi: Phys. Rev. B **64**, R121406 (2001)
64. Q. Guo, E.M. Williams: Surf. Sci. **433–435**, 322 (1999)
65. H. Uetsuka, A. Sasahara, A. Yamakata, H. Onishi, J. Phys. Chem. B in press
66. H. Uetsuka, A. Sasahara, H. Onishi, in preparation
67. A. Sasahara, H. Uetsuka, H. Onishi: J. Phys. Chem. B **105**, 1 (2001)
68. A. Sasahara, H. Uetsuka, H. Onishi: Appl. Phys. A **72**, S101 (2001)
69. A. Sasahara, H. Uetsuka, H. Onishi: Surf. Sci. **481**, L437 (2001)
70. A. Sasahara, H. Uetsuka, H. Onishi: Appl. Surf. Sci. **188**, 265 (2002)
71. P. Sawunyama, L. Jiang, A. Fujishima, K. Hashimoto: J. Phys. Chem. B **101**, 11000 (1997)
72. L. Patthey, H. Rensmo, P. Persson, K. Westermark, L. Vayssieres, A. Stashans, Å. Petersson, P.A. Brühwiler, H. Siegbahn, S. Lunell, N. Mårtensson: J. Chem. Phys. **110**, 5913 (1999)
73. K. Komiyama, M. Gu: Appl. Surf. Sci. **120**, 125 (1997)
74. S. Suzuki, Y. Yamaguchi, H. Onishi, T. Sasaki, K. Fukui, Y. Iwasawa: J. Chem. Soc. Faraday Trans. **94**, 161 (1998)
75. S. Suzuki, H. Onishi, T. Sasaki, K. Fukui, Y. Iwasawa: Catal. Lett. **54**, 177 (1998)
76. E. Soria, I. Colera, E. Roman, E.M. Williams, J.L. de Segovia: Surf. Sci. **451**, 188 (2000)
77. L. Gamble, M.A. Henderson, C.T. Campbell: J. Phys. Chem. B **102**, 4536 (1998)
78. M. Alam, M.A. Henderson, P.D. Kaviratna, G.S. Herman, C.H.F. Peden: J. Phys. Chem. B **102**, 111 (1998)
79. B.E. Hayden, A. King, M.A. Newton: Chem. Phys. Lett. **269**, 485 (1997)
80. R.A. Bennett, M.A. Newton, R.D. Smith, M. Bowker, J. Evans: Surf. Sci. **487**, 223 (2001)
81. J. Evans, B.E. Hayden, M.A. Newton: Surf. Sci. **462**, 169 (2000)
82. G.A. Rizzi, A. Magrin, G. Granozzi: Phys. Chem. Chem. Phys. **1**, 709 (1999)
83. S. Suzuki, Y. Yamaguchi, H. Onishi, K. Fukui, T. Sasaki, Y. Iwasawa: Catal. Lett. **50**, 117 (1998)

84. S. Suzuki, K. Fukui, H. Onishi, Y. Iwasawa: Phys. Rev. Lett. **84**, 2156 (2000)
85. D.A. Bonnell: Prog. Surf. Sci. **57**, 187 (1998)
86. E.L.D. Hebenstreit, W. Hebenstreit, U. Diebold: Surf. Sci. **461**, 87 (2000)
87. C.T. Campbell: J. Chem. Soc. Faraday Trans. **92**, 1435 (1996)
88. X. Lai, T. P St. Clair., M. Valden, D.W. Goodman: Prog. Surf. Sci. **59**, 25 (1998)
89. M. Bäumer, H.J. Freund: Prog. Surf. Sci. **61**, 127 (1999)
90. Y. Matsumoto, M. Murakami, T. Shono, T. Hasegawa, T. Fukumura, M. Kawasaki, P. Ahmet, T. Chikyow, S. Koshihara, H. Koinuma: Science **291**, 854 (2001)
91. M.A. Henderson: Surf. Sci. **355**, 151 (1996)
92. C.D. Bain: J. Chem. Soc. Faraday Trans. **91**, 1281 (1995)
93. L.J. Richter, T.P. Petralli-Mallow, J.C. Stephenson: Opt. Lett. **23**, 1594 (1998)
94. T. Ishibashi, H. Onishi: Appl. Phys. Lett. **81**, 1338 (2002)

6 Self-Assembled Monolayers for Molecular Nanoelectronics

Takao Ishida

Summary. In this chapter, we describe recent progress related to organic self-assembled monolayers (SAMs), ranging from the basic properties of SAMs to their application in molecular nano-electronics. The molecular arrangements and chemical properties of organosulfur SAMs on Au surface have been characterized using surface sensitive techniques, such as scanning probe microscopy, which have revealed that the organic molecules in the SAM were well ordered and densely packed. Also, nanoscale pattern formation in SAMs has been demonstrated using several self-organized methods, such as fill-in or co-adsorption techniques. The electrical conduction through SAMs was examined by scanning probe microscopy and using other other nanogap electrodes. In some conjugated molecules, a large negative differential resistance and memory device operation were reported. These interesting physical phenomena will provide us with ideas for the realization of nanoscale molecular devices.

6.1 Introduction

Organic molecules can be synthesized with unique properties that could be used to promote their self-assembly with one another and on specific surfaces, and to perform functions that could allow electronic-device operations. New conductive conjugated molecular wires have been found or synthesized for the purpose of constructing molecular devices. Self-assembled monolayers (SAMs) (Fig. 6.1) [1] provide a convenient technique to fix these functionalized organic molecules on suitable metal or semiconductor substrates. In particular, SAMs made from organosulfur compounds on an Au surface have been utilized for demonstration of nanomolecular electronics devices [2,3], because of the ease of formation of Au–S chemical bonds. Also, the fabrication of nanoscale structures using SAMs has attracted much attention; it is possible to form device-like structures such as nanodots [4] and nanowires [5]. For future application in nanoscale molecular devices, it is essential to explore new methods of using self-organization to obtain smaller nanoscale patterns and control the nanoscale structure. Here we describe recent progress in SAMs, mainly in relation to recent work in molecular nanoelectronics.

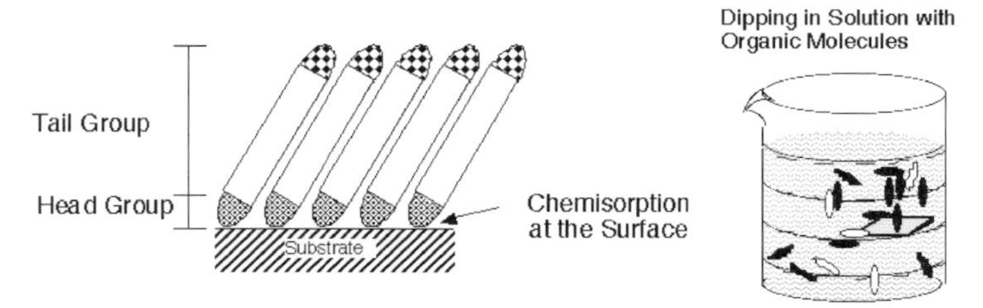

Fig. 6.1. Schematic drawing of SAMs

6.2 Basic Properties of Organosulfur SAMs on Au Surfaces

First, we describe the basic properties of organosulfur SAMs formed on Au surfaces. It is well known that organosulfur compounds react with Au surface. Taniguchi et al. observed the spontaneous formation of pyridine disulfide SAMs on an Au surface in 1982 [2]. Nuzzo and Allara reported the formation of alkanethiol SAMs in 1983 [3]. Typically, alkanethiols or dialkyl disulfides are utilized for organosulfur SAM formation. Dialkyl monosulfides can be also used for the same purposes [6–10]. However, the molecular arrangement and the strength of chemical bonding are different from those of alkanethiol or dialkyl disulfide SAMs. Typically, the Au(111) surface has been used for SAM formation because of the ease of preparation and stability of the crystal face. For other Au crystal faces, some studies have been performed, for example of SAMs on Au(100) [11]. However, the number of such studies is quite small.

(a) Monomer Structure (alkylthiolate)

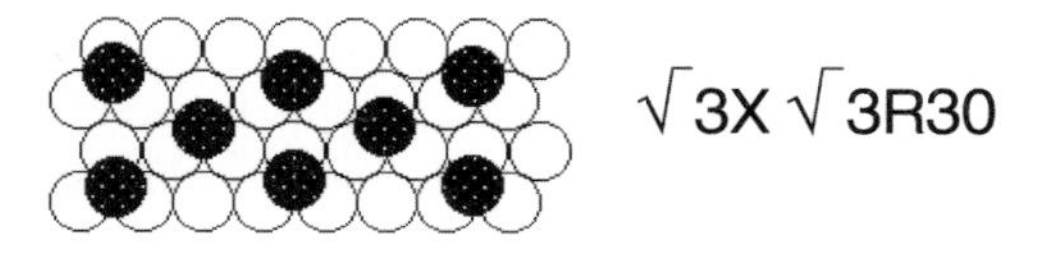

(b) Dimer Structure (disulfide)

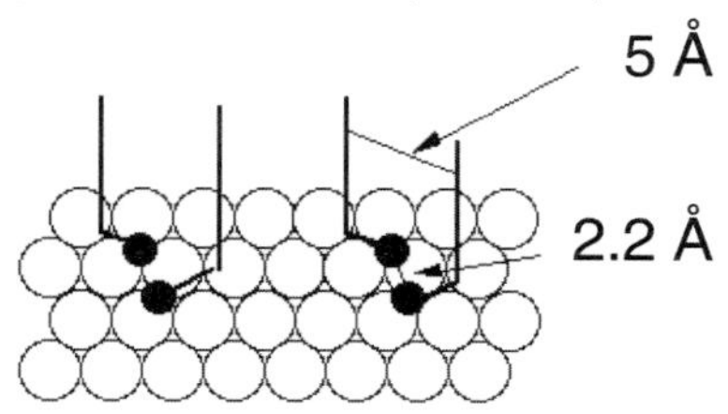

Fig. 6.2. Schematic drawing of alkanethiol SAMs: (**a**) monomer thiolate [11]; (**b**) dimer (disulfide) structure [13]

The molecular arrangements of alkanethiol SAMs, especially on the Au(111) surface, have been investigated using electron diffraction techniques, which revealed that the alkanethiol molecules exhibited a $(\sqrt{3} \times \sqrt{3})$ R30° structure (Fig. 6.2a). Dialkyl disulfide molecules formed an identical $(\sqrt{3} \times \sqrt{3})$ R30° structure to that of the alkanethiol SAMs [11]. Infrared (IR) measurements showed that alkyl chains longer than $n > 5$ were tilted about 30° from the surface normal [12]. Formerly, it was believed that both alkanethiols and dialkyl disulfides changed into alkylthiolates after the cleavage of the S–H bond or S–S bond. In this case, it was believed that the sulfur atoms of alkylthiolates were located at the 3-fold-hollow sites of the Au(111) surface. However, in 1994, Fenter et al. reported that such organosulfur compounds were present as a dimer structure (Fig. 6.2b), with almost the same structure as the dialkyl disulfides) [13]. However, although some studies appeared to show evidence of the presence of the dimer [14,15], no direct evidence of the dimer structure has been obtained yet [16–18]. Interestingly, Hayashi et al. recently proposed that the sulfur atoms of alkylthiolates are located not at the 3-fold-hollow sites but on the bridge sites of the Au (111) surface on the basis of high-resolution electron energy loss spectroscopy (HR-EELS) measuremnts and theoretical calculations [19]. Since such new findings appeared, the real adsorption sites and states of alkylthiolates have been in question.

Another class of organosulfur compound SAMs, dialkyl monosulfide SAMs, have also been investigated [6–10]. Zhong and Porter have studied the adsorption reaction of n-dibutyl sulfide on gold and concluded that this compound adsorbed with C–S cleavage in one of the side chains and formed an alkylthiolate SAM, identical to a SAM prepared from an alkanethiol (Fig. 6.3) [7]. However, recent studies have revealed that dialkyl monosulfides adsorb on Au intact without C–S cleavage and form less densely packed and less ordered SAMs compared with alkanethiol SAMs [10]. We found a poor reproducibility of the adsorption kinetics of monosulfide molecules onto an Au surface [9].

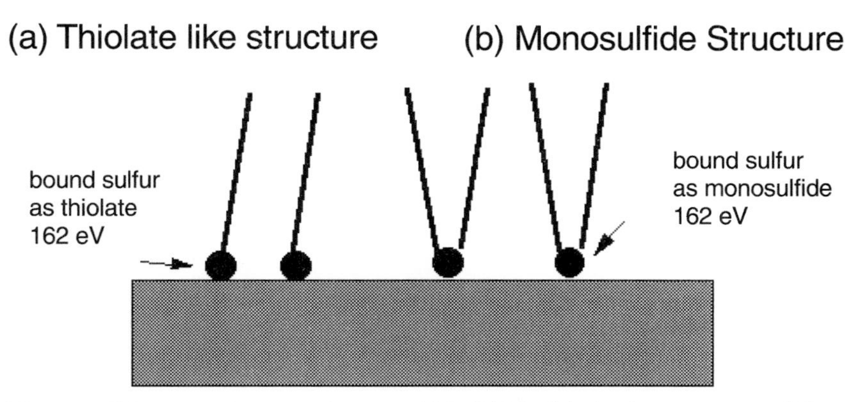

Fig. 6.3. Schematic drawing of monosulfide SAMs: (**a**) thiolate structure; (**b**) monosulfide structure

The poor reproducibility also makes the surface reaction of monosulfide molecules look more complicated. Interestingly, recent theoretical calculations have predicted that the Au atoms have a decreased conduction at the junction, and other functional groups to form chemical bonds with the Au surface have been proposed [20]. Semonario et al. concluded that a metal/N–C contact can be reduced tunneling resistance at the metal/molecular interface. Thus, it is important to find a new functional group to form a stable, high conductance metal/molecular contact, in place of the metal–S contact.

6.3 STM Observation of Organosulfur SAMs

Scanning probe microscopy (SPM)is a powerful technique to observe molecular arrangements, because SPM can visualize molecules in real space. In 1991, the surface topography of alkanethiol SAMs was investigated with scanning tunneling microscopy (STM) for the first time [21]. In that study, the presence of many depressions in the SAM surface was revealed [21]. In 1994, three groups succeeded in the direct observation of alkylthiolates on the Au(111) surface [22–24]. There exist two kinds of molecular lattice, i.e. the typical $(\sqrt{3} \times \sqrt{3})$ R30° structure and a C(4×2) superstructure [22–24]. These two structures coexist at the SAM surface (Fig. 6.4). It has been suggested that the C(4×2) structure originates from a twist about the alkyl chain axis. Interestingly, Touzov and Gorman observed a tip-induced transformation of

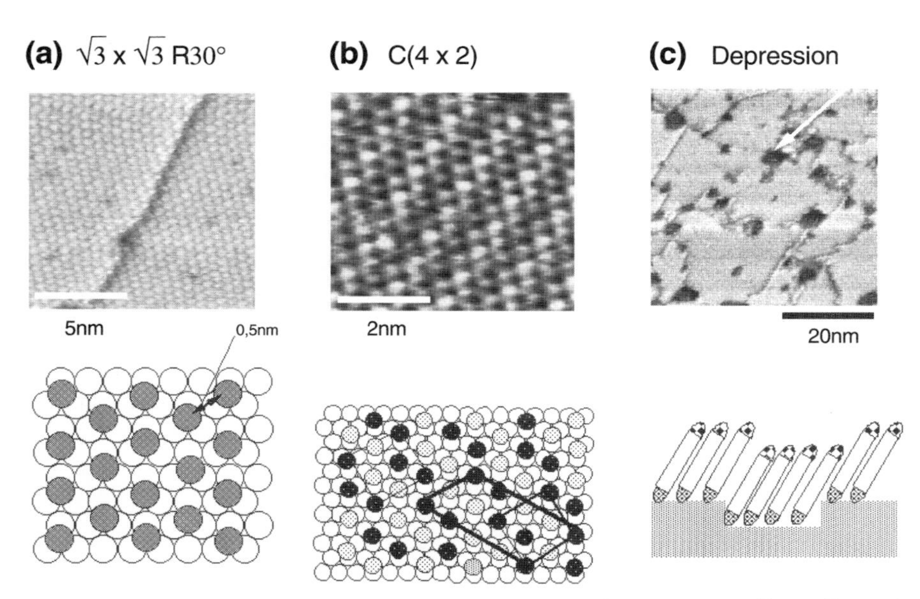

Fig. 6.4. STM images of typical alkanethiol SAMs [22–24]: (**a**) $\sqrt{3} \times \sqrt{3}$ R30° structure; (**b**) C(4×2) structure; (**c**) $100 \times 100\,\text{nm}^2$ image. Typical depressions are seen

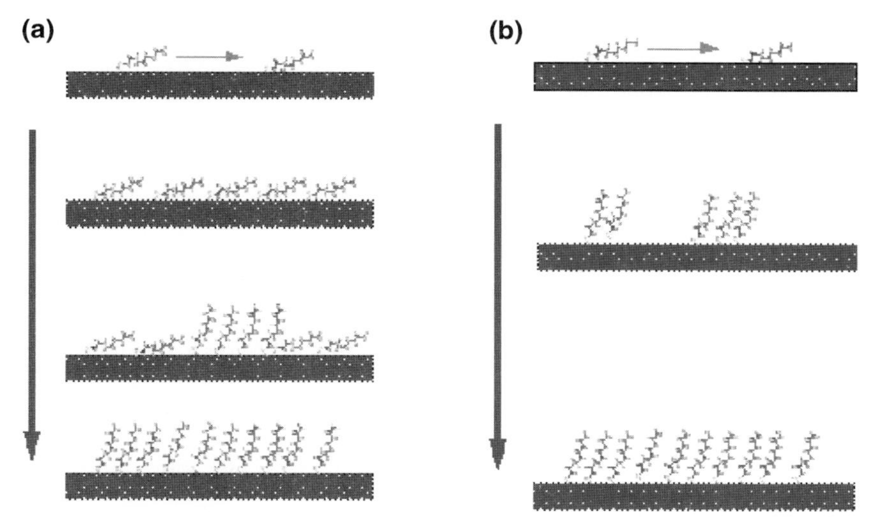

Fig. 6.5. Schematic drawing of the initial stage of alkanethiol SAM growth: (**a**) in vacuum or low-concentration solution [30,31]; (**b**) in a higher-concentration solution [32]

the molecular arrangement in alkanethiolates using STM [25]. In this case, even a lower tunneling current of 6 pA induced an irreversible reconstruction from the $(\sqrt{3} \times \sqrt{3})$ R30° structure to a C(4 × 2) superlattice. We found that the appearance of the molecular packing structures changed depending on the tip-surface distance when noncontact atomic force microscopy (AFM) was used [26]. Formerly, it was believed that such C(4 × 2) superlattices and typical $(\sqrt{3} \times \sqrt{3})$ R30° structures were the final phases of alkylthiolates on the Au(111) surface. Recently, Noh and Hara found a new stable phase of alkanethiol SAMs [27]. These authors reported that the C(4 × 2) superlattice gradually changed into a $6 \times \sqrt{3}$ phase over a period of 6 months. In the case of alkanethiol SAMs on Au(100), C(2 × 8) structures with a (1 × 4) arrangement of missing rows were observed by STM [28,29].

Adsorption processes of alkanethiols onto Au(111) surfaces have also been investigated by SPM. Poirier et al. observed vapor phase SAM formation. They found that in the initial stage of SAM growth, the alkyl chains were arranged to parallel the Au surface [30]. In the case of liquid phase SAM growth, the adsorption process depends on the concentration. At low concentration (0.001 mM), Yamada and Uosaki found that the alkylthiolates aggregated via a process of at least three steps [31]. However, at a higher concentration (0.01 mM), lying-down phases were not observed, i.e. two-step growth was confirmed (Fig. 6.5) [32,33].

Meanwhile, recently, STM observations of conjugated molecular SAMs have been reported, because such conductive conjugated molecules will be useful for future molecular-electronics applications. Highly ordered SAMs were successfully obtained in the case of oligo(phenylethynyl)arenethiol (Tour wire) SAMs [34,35]. Oligothiophenes are also attractive conjugated molecules. However, there exist few reports of STM observations of SAMs made from oligothiophenes [36,37]. In the case of oligothiophene derivatized thiols, both the thiophene rings and the thiol group can adsorb onto the Au surface directly. This reactivity may cause a difficulty in observation of ologothiophene derivatized thiol SAMs. Terphenyl (TP) and biphenyl (BP) derivatized thiol SAMs are other representative types of conjugated molecular SAMs [38–43]. Tao et al. [39] studied the effect of the methylene spacers of the conjugated molecules on the molecular arrangement on the Au surface. We have confirmed the influence of the methylene spacers on the molecular arrangement, as well as on monolayer electrical conduction [40]. Our STM images clearly indicated that the TP derivatives exhibited $(\sqrt{3} \times \sqrt{3})$ R30° structures when the methylene group was present (Fig. 6.6). On the other hand, in the case of a terphenylthiol without a methylene group (TP0), the molecularly resolved STM images could not be obtained [40]. In contrast, Kang et al. observed molecularly resolved STM images of 4-chlorobiphenylthiol SAM without a methylene group on the Au(111)surface [42].

Recently, we found that the adsorption process was strongly affected by the organic solvents in which the molecules were dissolved [43]. In methylene chloride solvent, an anisotropic nucleation along the $\langle 112 \rangle$ direction occurred in the initial stage of the TP0 SAM growth. At 1 min of immersion, the molecules were phase-separated into an ordered lying-down phase and a disordered phase. After more than 5 min of immersion, the ordered phases disappeared and formed larger striped patterns with a spacing of 8 nm. On the other hand, we found that the larger striped patterns did not form when we used ethanol as the solvent. In ethanol, the molecular packing of the TP0 molecules increased with the immersion time, while the domain size was small. Our data demonstrate that ethanol facilitated the formation of more densely packed TP0 SAMs than those formed in methylene chloride solvent. On the other hand, Tour wires without a methylene group can form highly ordered SAMs, unlike TP or BP SAMs. The reason for this discrepancy is not yet understood.

The SPM observation of supramolecules during SAM formation has also attracted attention recently [44–47]. For some supramolecules, since it is difficult to form an ordered structure in monocomponent supramolecular SAMs, a technique of insertion into a preassembled alkanethiol SAM is often used to fix isolated supramolecules. (The insertion process will be described in Sect. 6.4) For example, the observation of fixed dendrimer molecules embedded in alkanethiol SAMs has been reported [44]. We also succeeded in the fixation of Rotaxane molecules using tapping-mode AFM [45]. On the other hand, in some cases of monocomponent supramolecular SAMs, highly ordered structures can be observed using SPM techniques. Haga et al. reported SAM formation with molecules of a Ru complex molecules [46]. Cyclic voltammo-

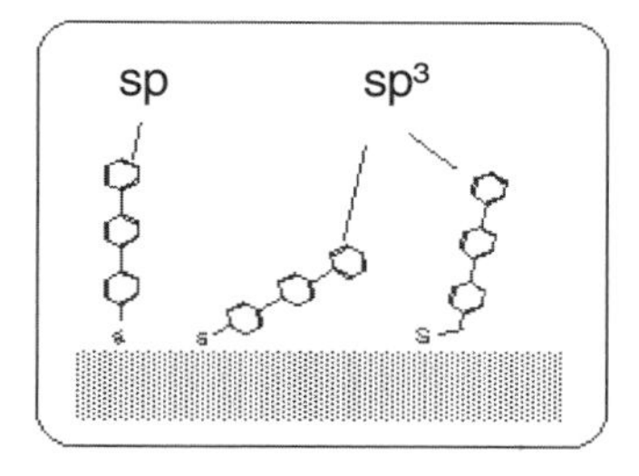

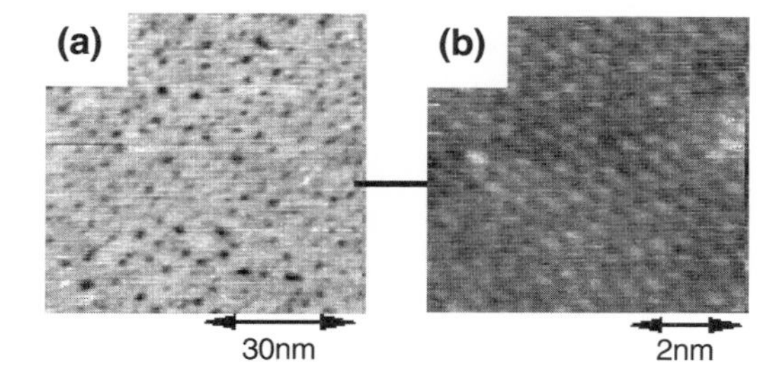

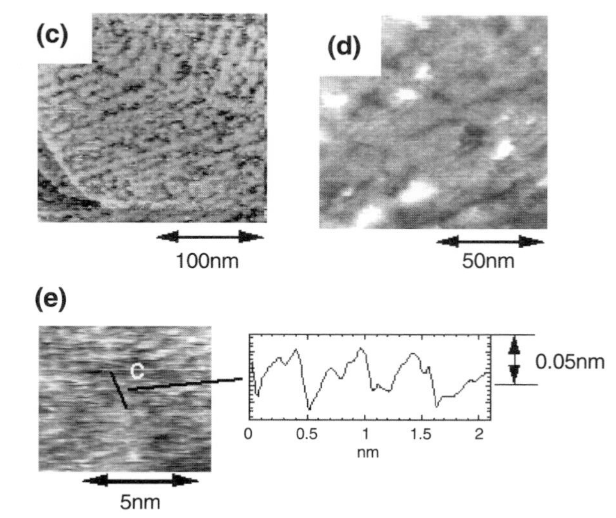

Fig. 6.6. STM images of typical TP SAMs [40,43]: (**a**) TP molecules with a methylene group between the sulfur and TP; (**b**) TP molecules without a methylene group (TP1) formed in methylene chloride solvent; (**c**) magnified image of (**b**); (**d**) TP molecules without a methylene group (TP0), formed in ethanol solvent; (**e**) magnified image of (**d**)

grams (CVs) for the reductive desorption of such a Ru complex indicated the existence of two different phases, i.e. a randomly oriented disordered phase and a well-ordered phase. We succeeded in the observation of the two phases by STM, and the STM images were consistent with the CV data [47].

6.4 Nanostructure Formation Using SAMs

Nanostructure formation using SAMs has also attracted much attention, because of its scientific importance and its potential application. Extensive efforts have been made to use SAMs as resist materials for electron beams [48]. Another technique, known as the microcontact printing method (Fig. 6.7a), utilizes a patterned elastomeric rubber as a stamp and an alkanethiol solution as an ink to directly form patterned features on an Au surface [49]. The microcontact printing method can produce patterns with line width less than 500 nm and is promising for industrial applications. However, for future application in nanoscale molecular devices, it is necessary to explore other self-organization methods for forming molecular nanostructures.

The SPM lithography technique has been studied by many people from this viewpoint of nanoscale patterning (Fig. 6.7b) [50–54]. Meanwhile, phase separation of mixed SAMs has been investigated by SPM. First, Stranick et al. [4] and Tamada et al. [32] observed phase separation in a SAM made by the coadsorption technique (Fig. 6.7c). However, in the case of coadsorption, it seems to be difficult to determine the surface composition. On the other hand, other techniques for forming binary SAMs, such as using asymmetric disulfides [55] (Fig. 6.7d) and an insertion (fill-in) technique [40,41,56–59] have been demonstrated (Fig. 6.7e). These techniques have the advantage that the ratio of the two molecular components on the surface can be easily controlled by changing the insertion time or the concentration of the molecule.

The insertion process has been investigated with the help of STM to form nanoscale structures and to measure the electrical properties of the moleucles [40,41,56,57]. Bumm and coworkers succeeded in implanting conductive molecules into insulative alkanethiols [56,57] and evaluated the conductivity

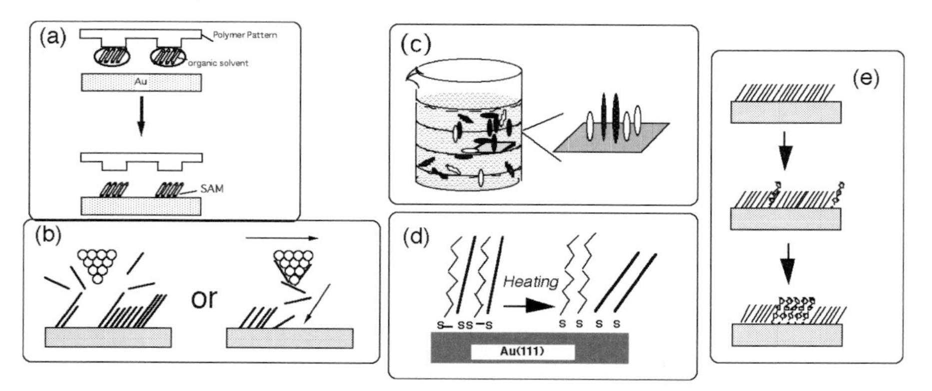

Fig. 6.7. Schematic drawing of methods of formation of nanostructures: (**a**) microcontact printing [49]; (**b**) SPM lithography [51–54]; (**c**) phase separation, coadsorption technique [4,32]; (**d**) phase separation, asymmetric disulfide technique [55]; (**e**) phase-separation, insertion (fill-in) technique [56–59]

of conjugated molecules. The electrical conduction of the molecules will be described in Sect. 6.5.

6.5 Electrical-Conduction Measurements of SAMs

There have appeared many interesting studies concerning electron transfer along the molecular axis. These studies have been performed both theoretically, and experimentally using SPM techniques [40,41,57,58,60–66], or the break junction technique [68], and a nanogap electrode [69,70] based upon the SAM technique. In this section, electrical-conduction studies of SAMs by SPM will be described.

First, Bumm et al. estimated the conductance of single conjugated molecules embedded in an insulative SAM film using STM [56,57]. These authors measured the height of molecules adsorbed on metal surface with the molecular axis almost parallel from the surface normal. We found various sizes of domains of the conjugated molecule [1,1′:4′,1″-terphenyl]-4-methanethiol (TP) in insulative alkanethiol SAMs, and evaluated both the vertical and lateral conductivities of TP domains using a conducting-disk model where intermolecular interaction may increase the electrical conduction (Fig. 6.8) [40,41]. We further the measured the electrical conduction of a series of conjugated molecules with phenyl rings embedded into alkanethiol SAMs, to investigate the effect of molecular structure on the electrical-conduction. In the case of conjugated molecules with one or three methylene groups between the sulfur and the phenyl ring, the measured height of the conjugated molecular domains depended on their lateral size, while a strong dependence was not observed in the case of a conjugated molecule without a methylene group. As a result of this analysis, one methylene group between the sulfur and the aromatic phenyl ring was found to be necessary to increase the vertical conduction of the molecular domains.

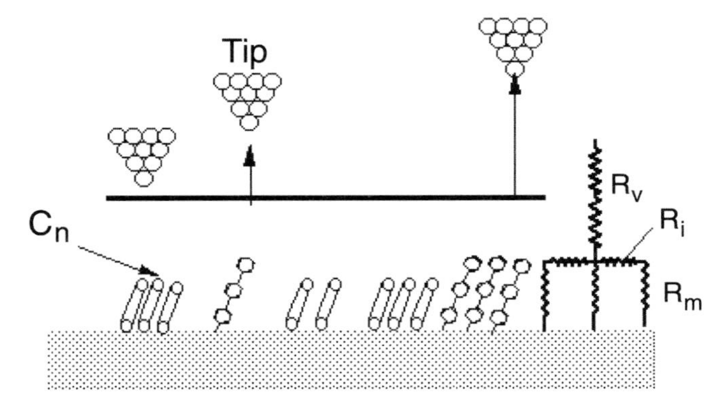

Fig. 6.8. Schematic drawing of electrical-conduction measurements of molecules in SAMs using STM [40,41,56,57]

However, using the STM technique, estimation of the electrical conduction of adsorbed molecules is quite complicated compared with other techniques such as using a break junction [68] or a nanogap electrode [69,70], where an electrode is directly attached to the molecules. While it is possible to estimate molecular conduction using scanning tunneling spectroscopy (STS), it is not so easy to minimize the effect of the tunneling gap between the STM tip and the molecules on the molecular-conductance measurements. We have proposed a new method for estimating molecular resistance based upon analysis of STM cross-sectional profiles [40,41]. In this method, to correlate the height difference with the resistance, an estimate of the tunneling resistance versus the z-displacement value ($\mathrm{d}R/\mathrm{d}z$) is required. We assumed that $\mathrm{d}R/\mathrm{d}z$ was constant and roughly estimated the molecular conduction. We explained qualitatively the change in the conduction of various molecular domains, but quantitative analysis was difficult.

On the other hand, atomic force microscopy (AFM) using a conductive cantilever is a candidate for direct measurement of electrical-conduction of organic monolayers such as SAMs. In current–voltage (I–V) measurements using AFM, the effects of the contact, or the value of the load between the AFM tip and the SAM surface, are important. For example, using conjugated molecules, Leatherman et al. measured the electrical conduction of carotenoid molecules embedded in alkanethiol SAMs, and obtained a single-molecule resistance of $4.2 \times 10^{10}\ \Omega$ [61]. They described the important effect of the applied load on the electrical conduction. Wold and Frisbie [62,63] recently measured the I–V curves of alkanethiol and benzylmercaptan SAMs systematically using conductive AFM. Their findings indicate that the current increased exponentially with the load value for insulative alkanethiol SAMs [62,63]. However, recently, Cui and coworkers found that gold particles directly connected via Au–S bonds assist smooth electrical transport in the case of alkyl chains [64]. On the other hand, for conjugated molecular SAMs, such a specific treatment to reduce the contact resistance is not necessary [65,66].

For conjugated molecular SAMs, detailed experimental data regarding the influence of the molecular structure on the electrical conduction are not yet available for measurements made with conductive AFM. We have measured the electrical-conduction of SAMs made from several kinds of phenylene oligomers with thiol groups, using AFM with a conductive tip, with the objective of understanding the effects of molecular structure as well as observing the influence of the load value on the electrical conduction (Fig. 6.9) [65,66]. We found that the load affected the electrical conduction of the terphenyl molecules differently from that of other conjugated SAMs. Since we had previously confirmed that the presence of methylene groups between the conjugated aromatic rings and the sulfur affects both the molecular arrangement and the electrical conduction, we used conjugated molecules with and without methylene groups between the sulfur and the aromatic rings, to investigate

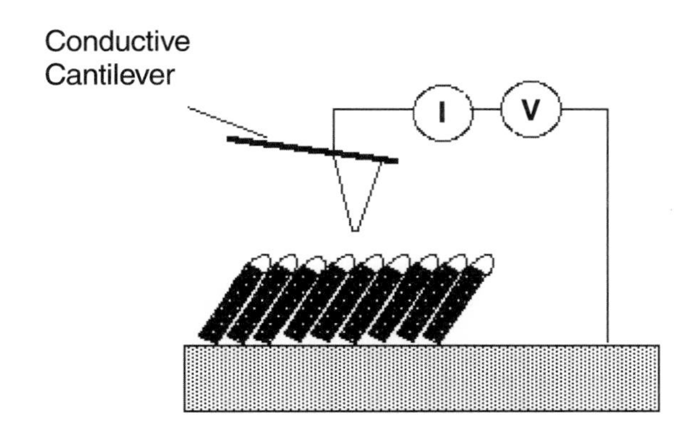

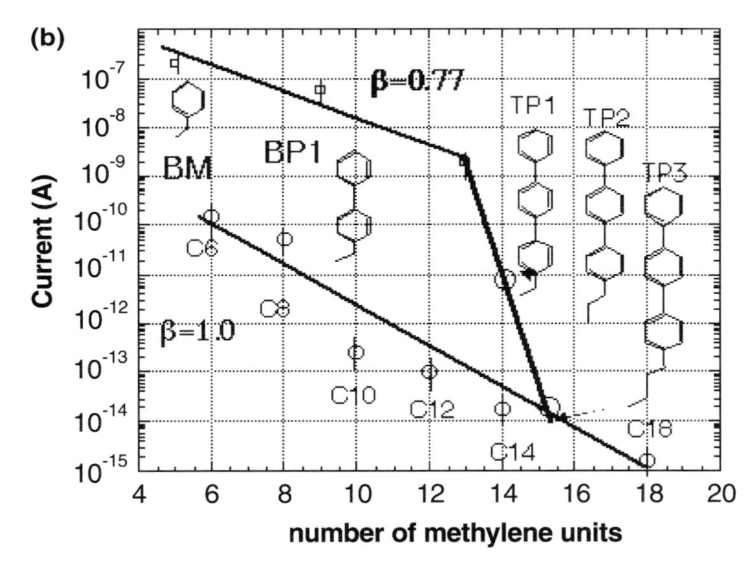

Fig. 6.9. (**a**) Schematic drawing of electrical-conduction measurement of molecules in a SAM using conductive AFM. (**b**) Dependence of current on the number of phenyl or methylene groups of conjugated molecular SAMs. All the currents were measured at +100 mV bias [65]

the effect of the distance between the metal and the molecule on the I–V curves. In addition, we found that direct contact of the conjugated ring with the Au surface further reduced the tunneling resistance [66].

Interestingly, the conductivities for conjugated molecules estimated using SPM are larger than those of undoped conductive organic materials [70].

This is one of the unanswered questions concerning the molecular conduction estimated by SPM techniques.

For a nanogap electrode, a large negative differential resistance was observed [71]. Memory device operation was reported using a similar nanogap electrode [72]. The detailed mechanisms are still unclear. In addition, the conformation, orientation, and arrangement of molecules in the electrode are not well known. However, these interesting physical phenomena will provide us with ideas for the realization of nanoscale molecular devices.

6.6 Stability of SAMs

The thermal stability of SAMs will be one of the important issues in future applications of SAMs in molecular devices, as well as in surface coatings, etc. The thermal stability of n-alkanethiol SAMs has been investigated [73,74]; for example, Delamarche et al. reported the effect of annealing dodecanethiol SAMs at around 100 °C [73]. During the annealing process, the depressions which are typically observed in alkanethiol SAMs almost disappeared owing to molecular diffusion, and then larger single-molecular domains without any depressions were formed. When the annealing was continued, the dodecanethiolates began to desorb from the Au surface. After 48 h of annealing at the same temperature, these dodecanethiolates were almost completely desorbed. A similar tendency was confirmed with a radiolabeling technique [74]. In the past, various strategies have been employed to increase the thermal stability of SAMs on Au surfaces, e.g. lateral polymerization [75], formation of a hydrogen bond network derived from amide groups [76], insertion of an underpotentially deposited metal layer between the molecules and the surface [77], and use of doubly binding molecules [78]. Another possible method to increase the thermal stability and avoid molecular decomposition is to utilize rigid molecules such as conjugated molecules, e.g. oligophenylenes; oligophenylene SAMs are expected to be rigid owing to the presence of phenyl rings. We have newly synthesized CF_3-terminated terphenyl derivative thiols to investigate the role of the terphenyl group as a spacer instead of methylene groups. For comparison, we also used tetradecanethiol (C_{14}) SAMs, the molecular length of which is identical to the that of CF_3-TP thiol. The effects of annealing these SAMs were examined to understand the thermal stability and annealing mechanism of conjugated molecular SAMs, by use of contact angle, STM, and X-ray photoelectron spectroscopy techniques [79]. Decreases in the contact angles of both of the CF_3-TP SAMs investigated were caused by annealing at 180 °C, while a similar decrease was observed at the lower temperature of 150 °C in the C_{14} SAMs. STM data revealed that the molecules were partially desorbed by the annealing process at the temperature where the contact angle of hexadecane began to decrease. These data demonstrate that the TP SAMs have higher thermal stability than the C_{14}

SAMs, and that the molecular-backbone structure is crucial in determinimg the thermal stability of the SAMs.

Schoenfisch and Pemberton investigated the stability of alkanethiol SAMs on Au and Ag surfaces in air [80], and observed rapid oxidation of sulfur in the monolayer. However, we confirmed using XPS that no oxidation of the sulfur occurred under our annealing condition [81]. If such a sulfur oxidation had occurred during annealing or exposure to air for several hours, we would have observed an oxidized-sulfur (sulfonate) peak at around 167 eV in the S(2p) region. But we detected only peaks in the range between 161 and 164 eV. Moreover, we could not detect any oxidized sulfur even after exposure to air for more than 10 days. For rapid oxidation to have occurred, there must have been uncharacterized factors such as ozone or UV, which do not seem to be present in the environment used in our studies at a sufficient level [81].

6.7 Conclusions and Future Expectations

Recent progress in the study of SAMs has been interesting and very rapid. These studies have demonstrated that SAMs are promising not only for application in molecular electronics but also for simple surface-coatings, tribological, and bionic applications. However, even for typical alkanethiol SAMs, the basic properties, especially the chemical state of the adsorbed molecules, are not well understood at the moment. Furthermore, the relationship between the molecular structure and local functions (electrical-conduction, etc.) is not clear. In the present chapter, we have focused on SAMs formed on Au surfaces. But, to fabricate nanoscale molecular devices, SAM formation on other suitable substrates need to be considered. Except for the studies on Au, SAMs on silicon have mainly been studied. The number of SAMs formed onto other substrates have been small, even though there exist many attractive substrates for molecular electronics, such as indium tin oxide and diamond. The metal/molecular electrical contact is also another important problem to be solved. Furthermore, even though many kinds of conjugated molecules have been synthesized, detailed experimental data regarding the influence of the molecular structure on the electrical conduction or on new physical phenomena on the nanometer scale are not yet available. To fabricate new molecular devices in the future, we really need to reconsider and overcome these serious problems.

Acknowledgments

The author acknowledges H. Tokumoto, W. Mizutani, H. Azehara, K. Miyake, S. Sasaki, K. Hiratani, Y. Nagawa, and K. Tamada (AIST); H. Fukushima, H. Takiguchi, and H. Hokari (SEIKO-EPSON); U. Akiba and M. Fujihira (Tokyo Institute of Technology); and M. Haga and M. Inoue (Chuo University) for their helpful experimental support and helpful discussions.

References

1. A. Ulman: *An Introduction to Ultrathin Organic Films From Langmuir–Blodgett to Self-Assembly* (Academic Press, New York 1991); A. Ulman: Chem. Rev. **96**, 1533 (1996); A. Ulman (ed.): *Thin Films: Self-Assembled Monolayers of Thiols* (Academic Press, New York 1998)
2. I. Taniguchi, K. Toyoshima, H. Yamaguchi, K. Yasukouchi: J. Chem. Soc., Chem. Commum 1032 (1982)
3. R.G. Nuzzo, D.L. Allara: J. Am. Chem. Soc. **105**, 4481 (1983)
4. S.J. Stranick, A.N. Parikh, D.L. Allara, P.S. Weiss: J. Phys. Chem. **98**, 11136 (1994)
5. T. Ishida, W. Mizutani, H. Tokumoto, H. Hokari, H. Azehara, M. Fujihira: Appl. Surf. Sci. **130–132**, 786 (1998)
6. E.B. Troughton, C.D. Bain, G.M. Whitesides, R.G. Nuzzo, D.L. Allara, M.D. Porter: Langmuir **4**, 365 (1988)
7. C.-J. Zhong, M.D. Porter: J. Am. Chem. Soc. **116**, 11616 (1994)
8. M.W.J. Beulen, B.-H. Huisman, P.A. van der Heijden, F.C.J.M. van Veggel, M.G. Simons, E.M.E.F. Biemond, P.J. Lande, D.N. Reinhoudt: Langmuir **12**, 6170 (1996)
9. H. Takiguchi, K. Sato, T. Ishida, K. Abe, K. Yase, K. Tamada: Langmuir **16**, 1703 (2000)
10. C.-J. Zhong, M.D. Porter: Langmuir **15**, 518 (1999)
11. L. Strong, G.M. Whiteside: Langmuir **4**, 546 (1988)
12. M.D. Porter, T.B. Bright, D.L. Allara: J. Am. Chem. Soc. **109**, 3559 (1987)
13. P. Fenter, A. Eberhardt, P. Eisenberger: Science **266**, 1216 (1994)
14. C. Zubragel, C. Deuper, F. Schneider, M. Neumann, M. Grunze, A. Schertel, C. Woll: Chem. Phys. Lett. **238**, 308 (1995)
15. N. Nishida, M. Hara, H. Sasabe, W. Knoll: Jpn. J. Appl. Phys. **35**, 5866 (1996)
16. D.G. Castner, K. Hinds, D.W. Grainger: Langmuir **12**, 5083 (1996)
17. A. Badia, L. Demers, L. Dickinson, F.G. Morin, R.B. Lennox, L. Reven: J. Am. Chem. Soc. **119**, 11104 (1997)
18. G.J. Kluth, C. Carraro, R. Maboudian: Phys. Rev. **B 59**, R10449 (1999)
19. T. Hayashi, Y. Morikawa, H. Nozoye: J. Chem. Phys. **114**, 7615 (2001)
20. J.M. Seminario, C.E. De La Cruz, P.A. Derosa: J. Am. Chem. Soc. **123**, 5616 (2001)
21. L. Haussling, B. Michel, H. Ringsdorf, H. Rohrer: Angew. Chem. Int. Ed. Engl. **30**, 569 (1991)
22. G.E. Poirier, M.J. Tarlov: Langmuir **10**, 2853 (1994)
23. E. Delamarche, B. Michel, C. Gerber, D. Anselmetti, H.-J. Guntherodt, H. Wolf, H. Ringsdorf: Langmuir **10**, 2869 (1994)
24. C. Schoenberger, J. Jorritsma, J.A.M. Sondag-Huethorst, L.G.J. Fokkink: J. Phys. Chem. **99**, 3259 (1995)
25. I. Touzov, C.B. Gorman: J. Phys. Chem. **B 101**, 5263 (1997)
26. T. Uchihashi, T. Ishida, M. Komiyama, M. Ashino, Y. Sugawara, W. Mizutani, K. Yokoyama, S. Morita, H. Tokumoto, M. Ishikawa: Appl. Surf. Sci. **157**, 244 (2000)
27. J. Noh, M. Hara: Langmuir **18**, 1953 (2002)
28. G.E. Poirier: J. Vac. Sci. Tehcnol. **B 14**, 1453 (1996)
29. R. Yamada, K. Uosaki: Langmuir **17**, 4148 (2001)

30. G.E. Poirier, E.D. Pylant: Science **272**, 1145 (1996)
31. R. Yamada, K. Uosaki: Langmuir **14**, 855 (1998)
32. K. Tamada, M. Hara, H. Sasabe, W. Knoll: Langmuir **13**, 1558 (1997)
33. M. Kawasaki, T. Sato, T. Tanaka, K. Takao: Langmuir **16**, 1719 (2000)
34. A.A. Dhirani, R.W. Zehner, R.P. Hsung, P. Guyot-Sionnest, L.R. Sita: J. Am. Chem. Soc. **118**, 3319 (1996)
35. G. Yang, Y. Qian, C. Engtrakul, L.R. Sita, G.-Y. Liu: J. Phys. Chem. **B 104**, 9059 (2000)
36. M.H. Dishner, J.C. Hemminger, F.J. Feher: Langmuir **12**, 6176 (1996)
37. T. Nakamura, H. Kondoh, M. Matsumoto, H. Nozoye: Langmuir **12**, 5977 (1996)
38. E. Sabatani, J. Cohne-Boulakia, M. Bruening, I. Rubinstein: Langmuir **9**, 2974 (1993)
39. Y.-T. Tao, C.-C. Wu, J.-Y. Eu, W.-L. Lin, K.-C. Wu, C. Chen: Langmuir **13**, 4018 (1997)
40. T. Ishida, W. Mizutani, U. Akiba, K. Umemura, A. Inoue, N. Choi, M. Fujihira, H. Tokumoto: J. Phys. Chem. **B 103**, 1686 (1999)
41. T. Ishida, W. Mizutani, N. Choi, U. Akiba, M. Fujihira, H. Tokumoto: J. Phys. Chem. **B 104**, 11680 (2000)
42. J.F. Kang, A. Ulman, S. Liao, R. Jordan, G. Yang, L.-Y. Liu: Langmuir **17**, 95 (2001)
43. T. Ishida, W. Mizutani, H. Azehara, F. Sato, N. Choi, U. Akiba, M. Fujihira, H. Tokumoto: Langmuir **17**, 7459 (2001)
44. A. Friggeri, H. Schoenherr, H.-J. van Manen, B.-H. Huisman, G.-J. Vancso, J. Huskens, F.C.J.M. van Veggel, D.N. Reinhoudt: Langmuir **16**, 7757 (2000)
45. H. Azehara, W. Mizutani, Y. Suzuki, T. Ishida, Y. Nagawa, H. Tokumoto, K. Hiratani: submitted to Langmuir
46. M. Haga, H.G. Hong, Y. Shiozawa, Y. Kawata, H. Monjushiro, T. Fukuo, R. Arakawa: Inorg. Chem. **39**, 4566 (2000)
47. K. Miyake, T. Ishida, S. Yasuda, H. Shigekawa, M. Inoue, M. Haga, S. Sasaki: in preparation
48. C.S. Dulcey, J.H. Georger, Jr., V. Krauthamer, D.A. Stenger, T.L. Fare, J.M. Calvert: Science **252**, 551 (1991)
49. A. Kumar, G.M. Whitesides: Science **263**, 60 (1994)
50. T.-Y. Kim, A.J. Bard: Langmuir **8**, 1096 (1992)
51. S. Xu, G.-Y. Liu: Langmuir **13**, 127 (1997)
52. J.K. Schor, R.M. Crooks: Langmuir **13**, 2323 (1997)
53. W. Mizutani, T. Ishida, H. Tokumoto: Langmuir **14**, 7197 (1998)
54. R.D. Piner, J. Zhu, F. Xu, S. Hong, C.A. Mirkin: Science **283**, 661 (1999)
55. T. Ishida, S.-I. Yamamoto, W. Mizutani, M. Motomatsu, H. Tokumoto, H. Hokari, H. Azehara, M. Fujihira: Langmuir **13**, 3261 (1997)
56. L.A. Bumm, J.J. Arnold, M.T. Cygan, T.D. Dunbar, L. Jones II, D.L. Allara, J.M. Tour, P.S. Weiss: Science **271**, 1705 (1996)
57. M.T. Cygan, T.D. Dunbar, J.J. Arnold, L.A. Bumm, N.F. Shedlock, T.P. Burgin, L. Jones II, D.L. Allara, J.M. Tour, P.S. Weiss: J. Am. Chem. Soc. **120**, 2721 (1998)
58. N. Nishida, M. Hara, H. Sasabe, W. Knoll: Jpn. J. Appl. Phys. **36**, 2379 (1997)
59. D. Hobara, T. Sasaki, S. Imabayashi, T. Kakiuchi: Langmuir **15**, 5073 (1999)
60. M. Salmeron, G. Neubauer, A. Folch, M. Tomitori, D.F. Ogletree, P. Sautet: Langmuir **8**, 3600 (1993)

61. G. Leatherman, E.N. Durantini, D. Gust, T.A. Moore, A.L. Moore, S. Stone, Z. Zhou, P. Lez, Y.Z. Liu, S.M. Lindsay: J. Phys. Chem. **B 103**, 4006 (1999)
62. J.D. Wold, C.D. Frisbie: J. Am. Chem. Soc. **122**, 2970 (2000)
63. J.D. Wold, C.D. Frisbie: J. Am. Chem. Soc. **123**, 5549 (2001)
64. X.D. Cui, A. Primak, X. Zarate, J. Tomfohr, O.F. Sankey, A.L. Moore, T.A. Moore, D. Gust, G, Harris, S.M. Lindsay: Science **294**, 571 (2001)
65. T. Ishida, W. Mizutani, Y. Aya, H. Ogiso, S. Sasaki, H. Tokumoto: J. Phys. Chem. **B 106**, 5886 (2002)
66. T. Ishida, W. Mizutani, H. Azehara, K. Miyake, Y. Aya, S. Sasaki, H. Tokumoto: Surf. Sci. **514**, 187 (2002)
67. H. Sagakuguchi, A. Hirai, F. Iwata, A. Sasaki, T. Nagamura, E. Kawata, S. Nakabayashi: Appl. Phys. Lett. **79**, 3708 (2001)
68. M.A. Reed, C. Zhou, C.J. Muller, T.P. Burgin, J.M. Tour: Science **278**, 252 (1997)
69. A.J. Epstein: MRS Bulletin **22**, 16 (1997)
70. C. Zhou, M.R. Deshpande, M.A. Reed, L. Jones II, J.M. Tour: Appl. Phys. Lett. **71**, 611 (1997)
71. J. Chen, M.A. Reed, A.M. Rawlett, J.M. Tour: Science **286**, 1550 (1999)
72. M.A. Reed, J. Chen, A.M. Rawlett, D.W. Price, J.M. Tour: Appl. Phys. Lett. **78**, 3735 (2001)
73. E. Delamarche, B. Michel, H. Kang, C. Gerber: Langmuir **10**, 4103 (1994)
74. J. Schlenoff, M. Li, H. Ly: J. Am. Chem. Soc. **117**, 12538 (1995)
75. P.N. Batchelder, T.H. Evans, T.L. Freeman, L. Haussling, H. Ringsdorf: J. Am. Chem. Soc. **116**, 1050 (1994)
76. R.S. Clegg, S.M. Reed, J.E. Hutchison: J. Am. Chem. Soc. **118**, 2486 (1996)
77. G.K. Jeannings, P.E. Laibinis: Langmuir: **12**, 6137 (1996)
78. W. Mizutani, M. Motomatsu, H. Tokumoto: Thin Solid Films **273**, 70 (1996)
79. T. Ishida, H. Fukushima, W. Mizutani, S. Miyashita, H. Ogiso, K. Ozaki, H. Tokumoto: Langmuir **18**, 83 (2002)
80. M.H. Schoenfisch, J.E. Pemberton: J. Am. Chem. Soc. **120**, 4502 (1998)
81. T. Ishida, N. Choi, W. Mizutani, H. Tokumoto, I. Kojima, H. Azehara, H. Hokari, U. Akiba, M. Fujihira: Langmuir **15**, 6799 (1999)

7 Supramolecular Chemistry on Solid Surfaces

Takuya Matsumoto, Tomoji Kawai, Takashi Yokoyama

Summary. The potential of molecular electronics is readily recognized from the viewpoint of miniaturization of microelectronic circuits, because organic molecules can satisfy the requirements of such circuits with their stable nanometer scale structure and with energy levels that can be tailored by chemical synthesis. Investigations on the electronic properties and functions of single molecules have made remarkable advances since the invention of scanning tunneling microscopy. However, such investigations have not directly led to the production of actual molecular-scale electronic devices, owing to the lack of effective technologies for the fabrication of nanoscale molecular structures. In response to these circumstances, supramolecular assembly is one good way to create nanoscale molecular circuits. In this chapter, recent investigations of self-assembled nanostructures on solid surfaces created via supramolecular interaction are presented. After the history and the present status of the field of molecular electronics are summarized in Sect. 7.1, the concept of the supramolecular approach to molecular nanostructures is briefly discussed in Sect. 7.2. Control of the size and shape of a supramolecular assembly by dipole–dipole interactions, introduced by asymmetric charge distributions around substitutins, is presented in Sect. 7.3. Highly selective supramolecular nanostructures formed by hydrogen bonds are discussed in Sect. 7.4. Finally, one-dimensional and ring-shaped nanostructures formed by π-stacking interactions are presented in Sect. 7.5.

7.1 Supramolecular Assembly in Molecular-Scale Electronics

The field of molecular electronics, first suggested by Amirav and Ratner [1], has attracted much attention in the last quarter-century. The device originally proposed was a rectifier based on electron transfer between donor and acceptor molecules. This concept became the headstream of a flow of macroscopic molecular devices such as soft plastic transistors, dye lasers and light-emitting diodes. In spite of their importance, these applications based on bulk organic materials are by-products rather than the core of Amirav and Ratner's concept [2], because bulk-based molecular devices exhibit not the functions of individual molecules but the band properties of crystalline or amorphous materials.

Molecular-scale electronics using the energy levels of a single molecule or several molecules is of great importance [3–5]. This is because the miniaturization of microelectronic circuits will soon require atomic- or molecular-scale components. Organic molecules would satisfy the needs of such circuits with their rigid, stable, well-defined, nanometer-scale structure and with energy levels that can be designed arbitrarily by chemical synthesis [6]. Recently, molecular-scale electronics has been considered from the viewpoint of biological and biomimetic applications. Molecules have an affinity to biological systems, and the network structure of self-assembled molecules is expected to open up a new paradigm for computing systems such as our brain [7].

Notwithstanding the great anticipation of the potential of molecular-scale electronics, fundamental investigation has been limited to liquid-phase systems for a long time. In this field, many experimental and theoretical studies of electron transfer between molecules have been performed [8]; however, such studies of isolated molecules do not provide a means for understanding electron transport through a single molecule connected to a metal electrode. The breakthrough beyond this limitation was brought about by the invention of scanning tunneling microscopy (STM), which enables us to establish electrical contact with individual molecules [9,10]. However, although STM gives fascinating results for the conductivity of a single molecule, such investigations have not directly led to actual molecular-scale electronic devices, owing to the lack of effective technologies for the fabrication of nanoscale molecular structures. Top-down methods such as lithography are not suitable for fabricating nanoscale molecular structures. On the other hand, the manipulation of individual molecules, the ultimate bottom-up process, is also not realistic, because of the extremely low production efficiency. For this reason, supramolecular assembly is expected to be a realistic approach to creating nanoscale molecular circuits [11].

7.2 Supramolecular Approach

Since supramolecular interactions have a moderate bonding energy, intermediate between the energies of covalent bonds and thermal excitation, they can form nanoscale structures while maintaining the nature of the individual molecules. A supramolecular approach starting from molecular building blocks can lead to a wide diversity and programmability that can be used to form nanoscale molecular structures, by use of selective and directional intermolecular interactions [11]. As shown in Fig. 7.1, when noncovalent intermolecular interactions such as hydrogen bonding are introduced into functional molecules, they result in the controlled formation of molecular nanostructures. So far, these structures have been obtained almost exclusively in crystals or in solution. To adapt these functional supramolecular structures to nanodevices, it will be necessary for the supramolecular structures to be supported at suitable positions on suitable substrates. Recent advances in high-resolution STM imaging allow one to directly determine the arrangement, configuration, and conformation of individual largish molecules on surfaces.

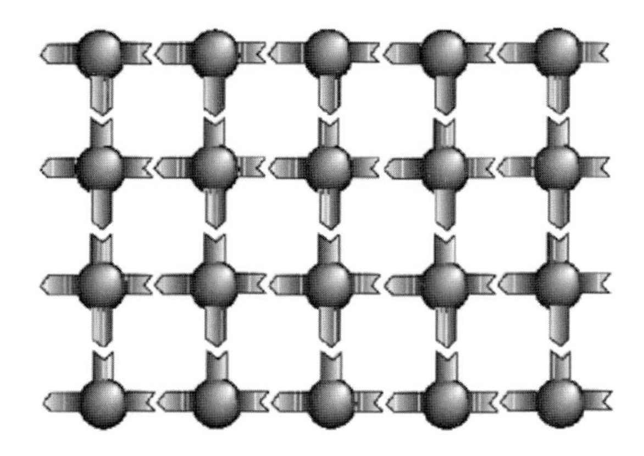

Fig. 7.1. Schematic image of a supramolecular structure. Individual molecules, denoted by *circles*, are connected by selective intermolecular interactions

7.3 Dipole–Dipole Interactions

The properties of organic molecules are drastically influenced by functional groups. For instance, benzene, a typical nonpolar molecule, is converted into water-soluble phenol by introduction of a hydroxyl group. Similarly, substituting only one functional group can strategically control a self-assembled structure.

High-resolution STM images of naphthalene ($C_{10}H_8$) chemisorbed on Pt(111) have been obtained as shown in Fig. 7.2 [12]. The naphthalene molecules appear as bi-lobed structures situated on (3×3) Pt lattice sites. The ordered overlayer is divided into small domains less than 50 Å in size with boundaries involving a shift of the molecules by an additional Pt lattice shift.

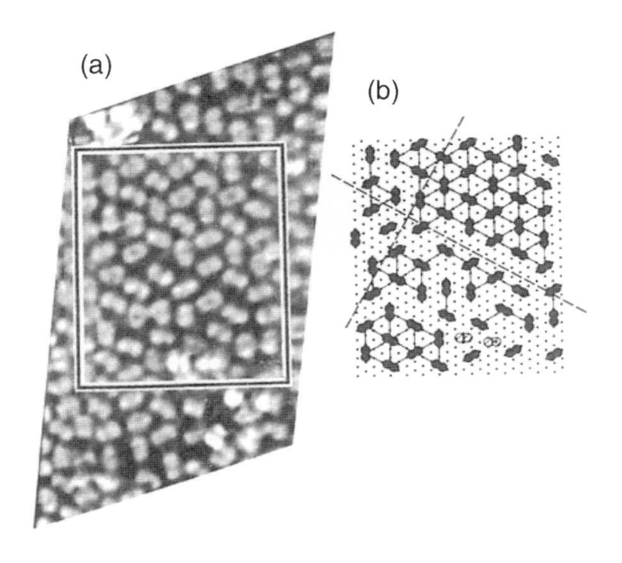

Fig. 7.2. (**a**) High-resolution STM image of ordered naphthalene on Pt(111) at room temperature. (**b**) Schematic diagram of the overlay of the Pt(111) lattice with molecular positions (from V.M. Hallmark et al., Phys. Rev. Lett. **66**, 48 (1991), reprinted with permission)

In a domain, the molecules exhibit three rotational orientations but their correlation is incomplete. Such a behavior of naphthalene molecules implies that the anisotoropy of the intermolecular interaction is not strong enough to overcome the sticking energy between the molecules and the surface. This situation is dramatically changed by the addition of a nitroxide group.

Schneider et al. have observed self-assembled clusters with a distinct size and structure using 1-nitronaphthalene (NN, structural formula in Fig. 7.3a) on a reconstructed Au(111) surface by using low-temperature STM [13]. As shown in Fig. 7.3b, most of the clusters formed on the Au(111) surface are

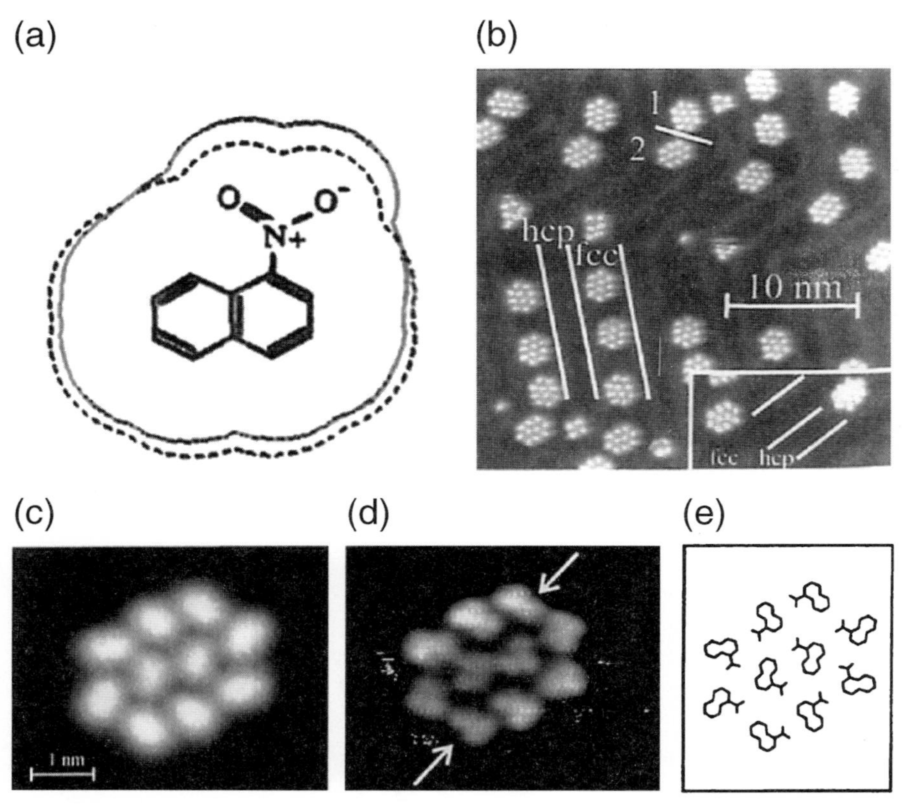

Fig. 7.3. (**a**) Structural formula of 1-nitronaphthalene (NN). The *dashed line* encloses the exclusion area induced by steric repulsion. The strength of the negative electrostatic potential is indicated by the distance of the *solid line* from the *dashed line*. (**b**) STM image at 50 K of a reconstructed Au(111) surface after 0.1 ML adsorption of NN. *Inset*, 0.2 ML NN at 10 K. Supramolecular decamers are formed between reconstructed rows of the Au(111) surface. (**c**) STM image of a decamer and (**d**) at high resolution. (**e**) Theoretical local-minimum structure for the observed decamer obtained from a force model (from M. Bohringer et al., Phys. Rev. Lett. **83**, 324 (1999), reprinted with permission)

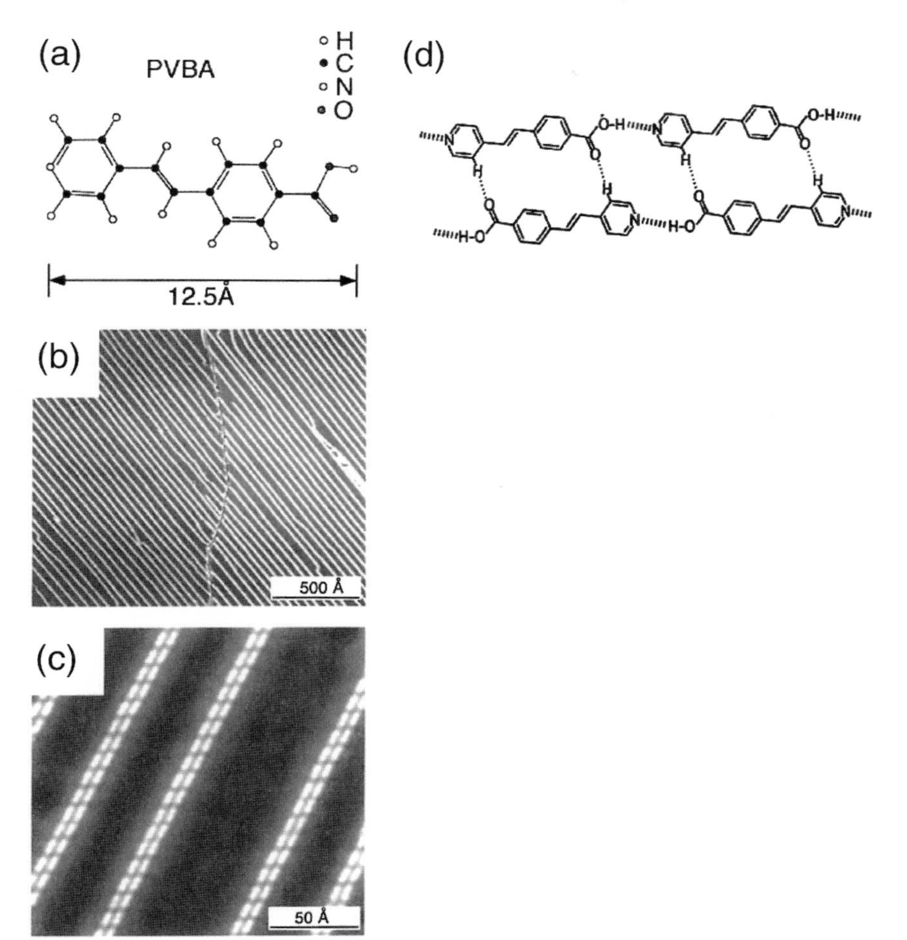

Fig. 7.4. (**a**) Structural formula of 4-[trans-2-(pyrid-4-yl-vinyl)]benzoic acid (PVBA). (**b**) STM image at 70 K of a one-dimensional supramolecular PVBA structure formed on an Ag(111) surface. (**c**) Image at high resolution. The structure is composed of coupled (twin) rows of PVBA molecules. (**d**) Structural model of the PBVA twin chain with OH...N and CH...OC hydrogen bonds (from J.V. Barth et al., Angew. Chem. Int. Ed. **39**, 1230 (2000), reprinted with permission)

composed of ten NN molecules, including an eight-molecule ring surrounding a two-molecule core. The structure and stability of the decamers are expected to be determined by specific intermolecular interactions that are related to the asymmetry of the molecular charge at the NO_2 group. This supramolecular structure has been confirmed by theoretical calculations, as presented in Fig. 7.3e, where the negatively charged oxygen atoms are attached to the hydrogen atoms opposite to the NO_2 group of a neighboring NN molecule.

A more complex intermolecular interaction has been observed in 4-[trans-2-(pyrid-4-yl-vinyl)]benzoic acid (PVBA, structural formula in Fig. 7.4a), which comprises a pyridyl group and a carboxylic acid group that provide head-to-tail hydrogen bonding. Barth et al. have demonstrated that PVBA molecules form highly regular, one-dimensional suprapmolecular arrangements on Au and Ag(111) surfaces [14]. As shown in Fig. 7.4b,c, the low-temperature STM images show long twin chains that are stabilized by strong head-to-tail OH...N hydrogen bonds between PVBA end groups as shown in Fig. 7.4d.

Now that these surface-supported supramolecular structures have been directly observed using STM, further control of their size and shape should become the next step in realizing molecular nanodevices. Recently, Yokoyama et al. have demonstrated the selective assembly of supramolecular aggregates with controlled size and shape by modifying the substituent structures [15]. These authors used largish porphyrin molecules, based on 5,10,15,20-tetrakis-(3,5-di-tertiarybutylphenyl) porphyrin (H2-TBPP, structural formula in

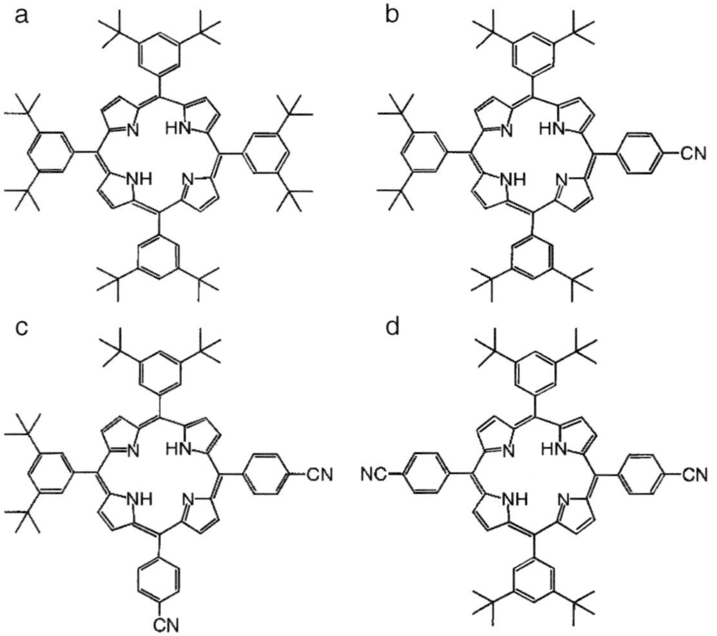

Fig. 7.5. Structural formulas of porphyrin molecules:
(**a**) tetrakis-(di-tertiarybutylphenyl)porphyrin (H2-TBPP);
(**b**) (cyanophenyl)-tris(di-tertiarybutylphenyl)porphyrin (CTBPP);
(**c**) cis-bis(cyanophenyl)-bis(di-tertiarybutylphenyl)porphyrin (cis-BCTBPP);
(**d**) trans-bis(cyanophenyl)-bis(di-tertiarybutylphenyl)porphyrin (cis-BCTBPP)
(from T. Yokoyama et al., Nature **413**, 619 (2001), reprinted with permission)

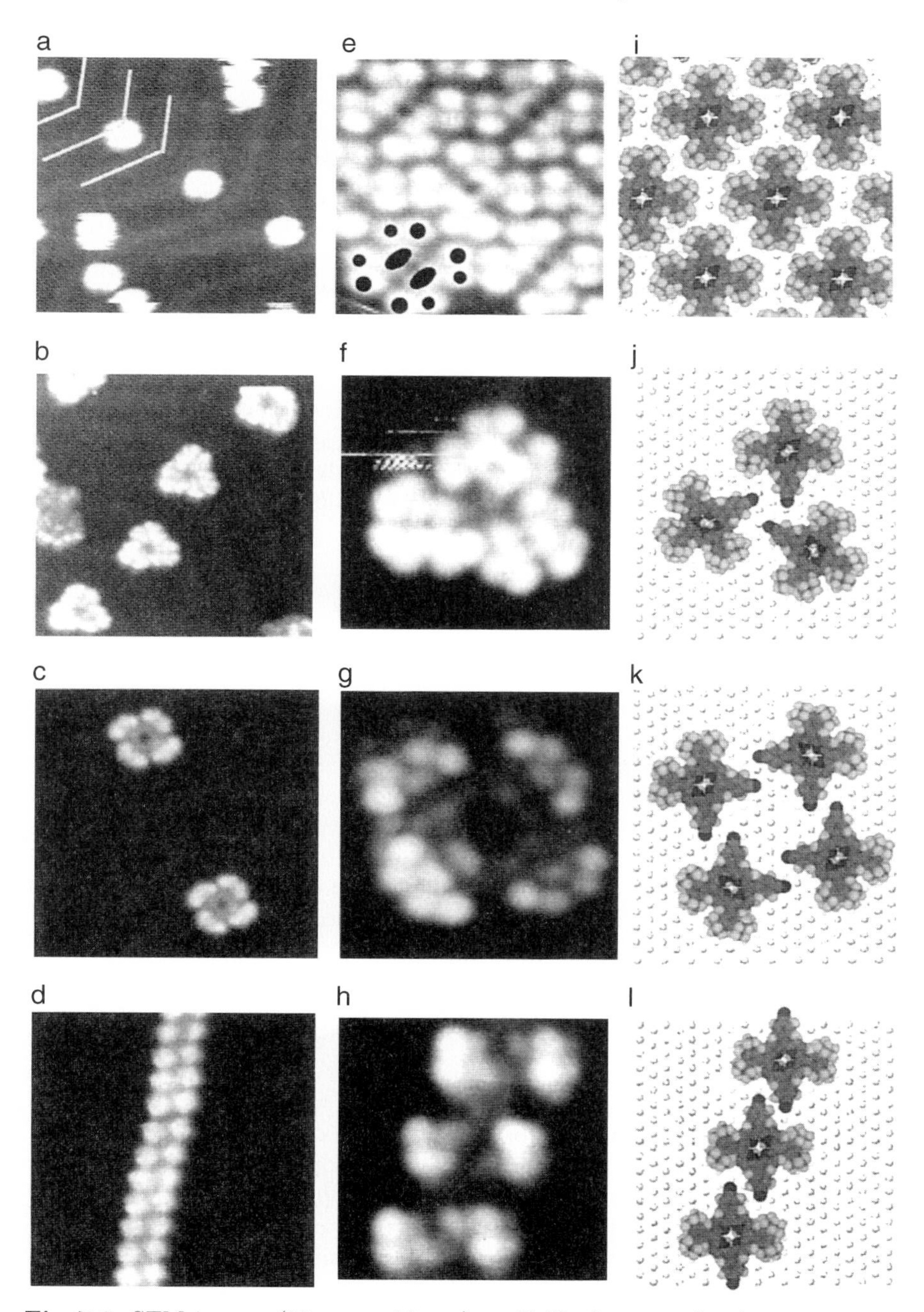

Fig. 7.6. STM images (20 nm × 20 nm) at 63 K of supramolecular aggregation induced by cyano groups on Au(111) surface: (**a**) H2-TBPP, (**b**) CTBPP, (**c**) cis-BCTBPP, (**d**) trans-BCTBPP. High-resolution STM images (5.3 nm × 5.3 nm) and the corresponding molecular models: (**e**), (**i**) H2-TBPP island; (**f**), (**j**) CTBPP trimer; (**g**), (**k**) cis-BCTBPP tetramer; (**h**), (**l**) trans-BCTBPP wire (from T. Yokoyama et al., Nature **413**, 619 (2001), reprinted with permission)

Fig. 7.5a) as building blocks. On the Au(111) surface, the STM image of an H2-TBPP thin film exhibits a simple close-packed arrangement as shown in Fig. 7.6, in which each single molecule consists of four paired lobes surrounding two oblong protrusions. This arrangement is that expected to result from simple van der Waals interactions [16].

An asymmetric charge distribution at the cyano (CN) group introduces dipole–dipole interactions between neighboring cyanophenyl substituents, and the resultant specific aggregation of cyanobenzene molecules has been predicted by theoretical calculations. To employ these specific interactions, one or two cyanophenyl substituents have been synthetically introduced into H2-TBPP as shown in Fig. 7.5. These molecules were selectively assembled into supramolecular trimer, tetramer, or extended wire-like structures, depending on the number and position of the cyanophenyl substituents. In these assembled structures, the cyano groups have an antiparallel or a cyclic configuration that is in good agreement with the theoretical predictions. Accordingly, these self-assembled structures are expected to be controlled by local intermolecular interactions even on surfaces.

7.4 Hydrogen Bonds

Functional groups consisting of hydrogen and anionic atoms such as oxygen, nitrogen, fluorine, or sulfur form hydrogen bonds. Their bonding energy is only one-tenth of that of a covalent bond but they are very important, owing to their presence in biological systems. For example, it is well known that deoxyribonucleic acid (DNA) has a double-stranded helical structure hybridized by specific hydrogen bonds between adenine and thymine, and between guanine and cytosine. This high selectivity originates from the structural rigidity of hydrogen bonds, where the shortest bond distance of the central hydrogen atom strongly restricts the geometry of the terminal anionic atoms.

Hydrogen-bonded rodlike polymeric nanostructures including calyx[4]arene **1** have been reported (Fig. 7.7) [17]. Combination of this compound with 5,5-diethylbarbituric acid (DEB) results in a self-assembled disklike nanostructure. This structure is "closed" and does not give polymers, because all hydrogen bonds are satisfied in a $1_3 \cdot 2_3$ supramolecule. On the other hand, in combination with calyx[4]arenedicyanurate **2**, a mismatching "open" form is obtained, which gives infinite rodlike stacks by noncovalent polymerization (Fig. 7.7) [18]. The nanostructure of $[1_3 \cdot 2_3]_n$ on a graphite surface was observed by tapping-mode scanning force microscopy, as shown in Fig. 7.8a–c. The sample was prepared by depositing a droplet of a dilute equimolar solution of **1** and **2** on a graphite surface. After evaporation of the solvent, perfectly aligned rodlike structures were observed. In comparison, $[1_3{\cdot}(\mathrm{DEB})_6]_n$ assemblies also show a rodlike structure (Fig. 7.8d,e) but this structure is not clearer than that of $[1_3 \cdot 2_3]_n$. This is because the rodlike assemblies are

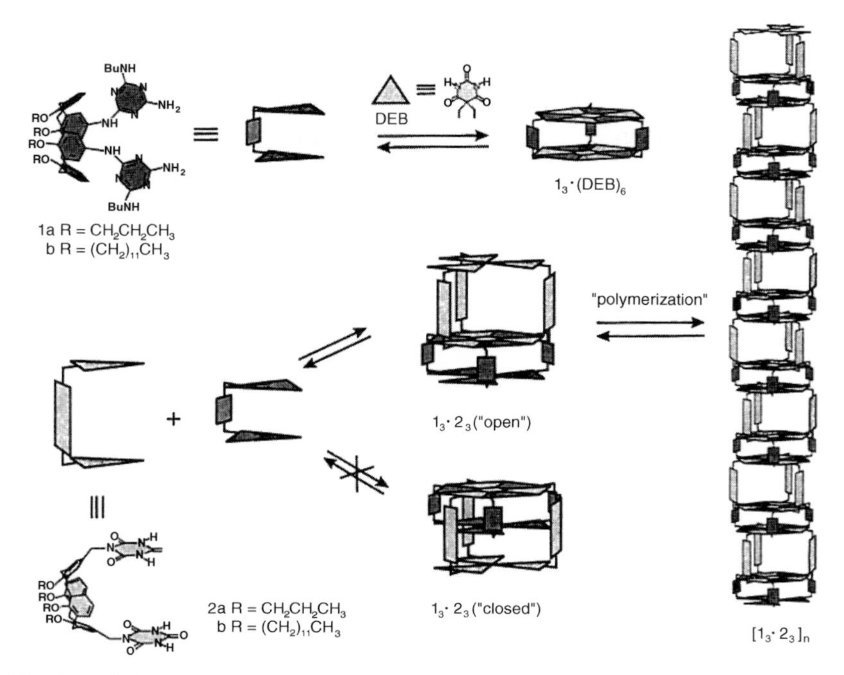

Fig. 7.7. Schematic representation of molecular components **1** and **2** and the rodlike nanostructures formed by hydorogen bonding (from H.-A. Klok et al., J. Am. Chem. Soc. **121**, 7154 (1999), reprinted with permission)

likely to be formed by solvophobic interactions. These structures have been analyzed according to the model presented in Fig. 7.8f.

Adenine, one of the DNA base molecules, forms various low-dimensional structures at low temperatures as shown in Fig. 7.9 [19]. According to theoretical calculations, five kinds of stable adenine dimers, (AAI–AAV in Fig. 7.9, linked by two hydrogen bonds, have been predicted. These dimers are the building blocks for more complicated self-assembled structures, as shown in Fig. 7.9. Once the most stable adenine dimers AAI have been formed, only the AAIV or AAV interaction is permitted for the interdimer bonding, because the molecular configuration of AAI prevents other types of linkage. The AAIV and AAV linkages result in a one-dimensional chain and a hexagonal structure, respectively.

7.5 π-Stacking Interactions

A great deal of interest has been directed toward aromatic compounds, since the delocalized π-electrons, which can be polarized and excited in the low-energy regime, exhibit electronic, magnetic, and optical functions. Crystals

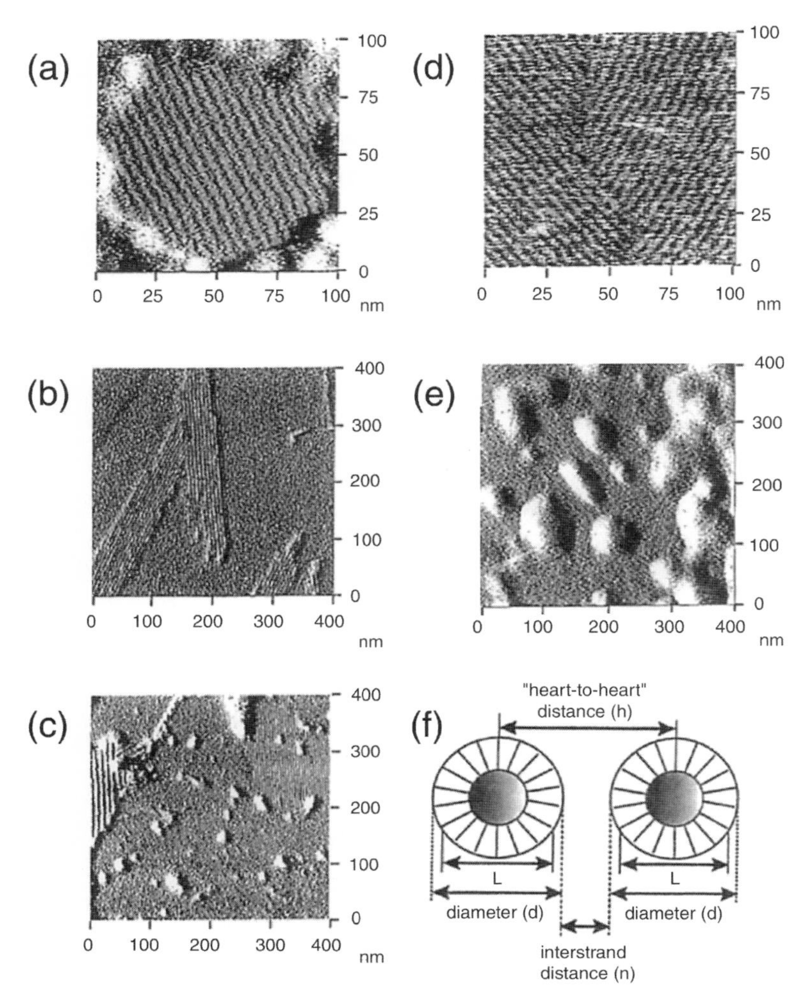

Fig. 7.8. Tapping-mode scanning force microscopy images of rodlike nanostructures: (**a**) $[1a_3 \cdot 2a_3]_n$(DMSO); (**b**) $[1b_3 \cdot 2a_3]_n$(THF); (**c**) $[1b_3 \cdot 2b_3]_n$(chloroform); (**d**) $[1a_3{\cdot}(\mathrm{DEB})_6]_n$(chloroform); (**e**) $[1b_3{\cdot}(\mathrm{DEB})_6]_n$(chloroform); (**f**) schematic front view of a possible structure model (from H.-A. Klok et al., J. Am. Chem. Soc. **121**, 7154 (1999), reprinted with permission)

of aromatic molecules show striking phenomena such as superconductivity, ferromagnetism, and nonlinear optical properties. The stacking of aromatic rings is a key factor governing these properties. However, knowledge about how to design functional materials of this kind cannot be applied to nanoscale devices fabricated on a substrate, since π-stacking is seriously affected by molecule–surface interactions. Accordingly, investigation of the π-stacking

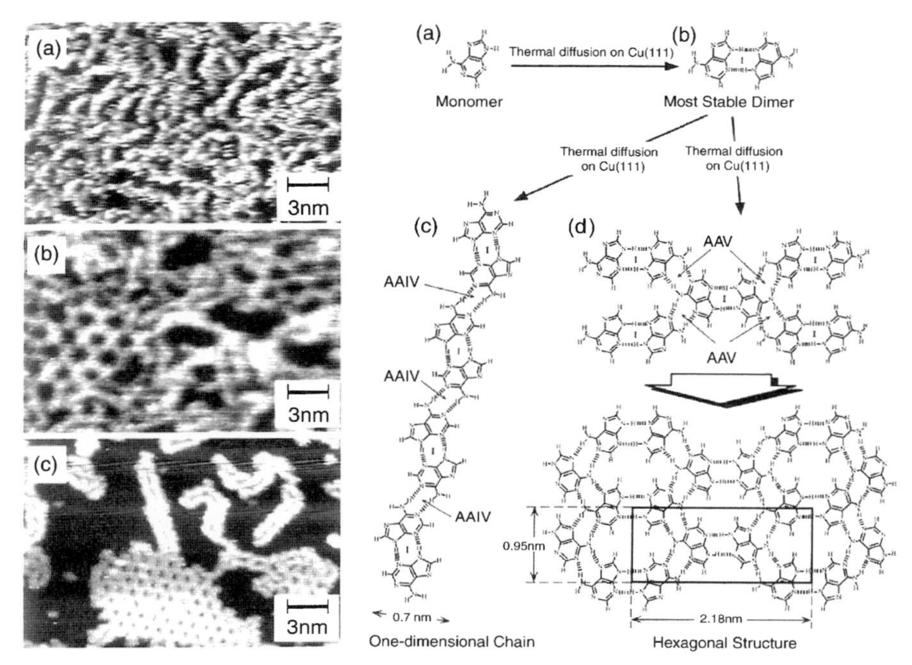

Fig. 7.9. Left panel: STM images of self-assembled adenine molecules. (**a**) Coverage $\theta = 1.0$ ML at room temperature. (**b**) $\theta = 1.0$ ML at 70 K. (**c**) $\theta = 0.3$ ML at 70 K. Right panel: the possible mechanism of the creation of self-assembled nanostructures of adenine molecules. (**a**) Adenine molecule (monomer); (**b**) adenine dimer (AAI); (**c**) one-dimensional chain linked by AAIV; (d) hexagonal structure linked by AAV (from M. Furukawa et al., Surf. Sci. **445**, 1 (2000), reprinted with permission)

at surfaces or interfaces is of great importance in the attempt to achieve nanoscale molecular devices.

Phthalocyanines are attractive molecules with a wide variety of applications owing to their narrow HOMO–LUMO gap, diversity of center-coordinated metal atoms, and high stability during thermal and chemical processing. The initial growth of Cu-phthalocyanine (CuPc) on highly oriented pyrolitic graphites (HOPG) surfaces has been well investigated by electron diffraction and STM. These studies have revealed that the CuPc layers initially grow as flat-lying molecules and that π-stacking between molecules is not present in the first layer. In fact, STM images of CuPc monolayers show close-packed flat-lying molecules on the HOPG surface [20]. In contrast, a monolayer of zinc phthalocyanine (ZnPc) shows a one-dimensional structure consisting of π-stacked molecules on an HOPG surface [21]. The reason why this structure can be formed is that the central Zn atom generates a polarization of the π-electrons since the electronegativity of Zn is larger than that of Cu. When the ZnPc coverage is under 1 ML, twinned row

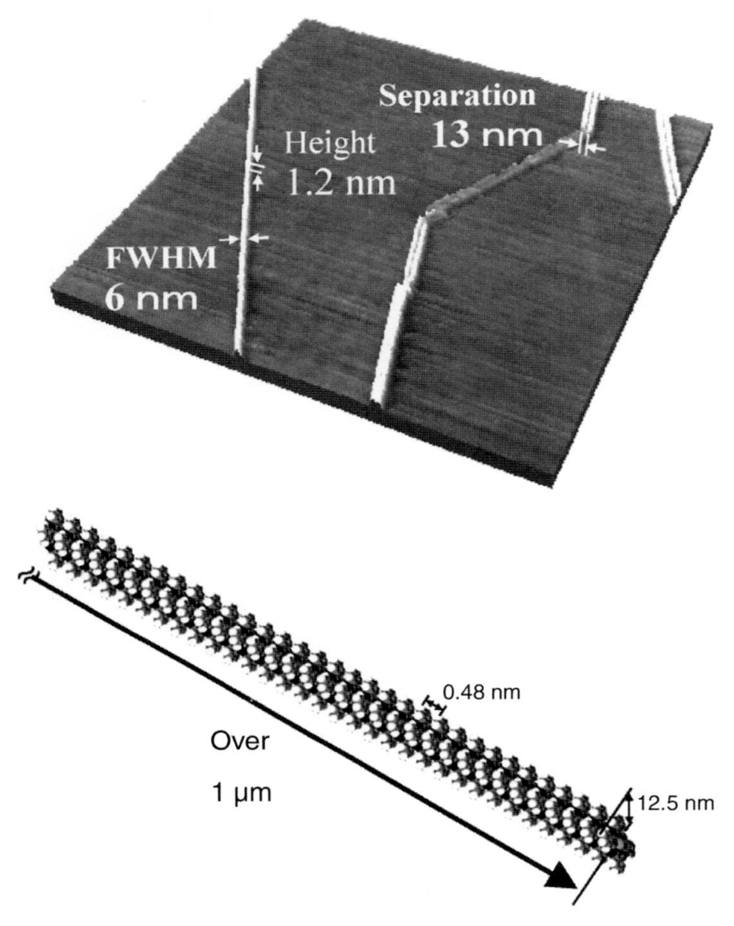

Fig. 7.10. STM image of one-dimensional nanostructures of zinc phthalocyanine on highly oriented pyrolic graphite at room temperature

structures with three-fold symmetry have been reported. Additional deposition over 1 ML results in long, self-assembled, one-dimensional (1D) wires (Fig. 7.10). The observed height, length and width of these 1D structures are 1.2 nm, more than several micrometers and less than 6 nm, respectively. The height of 1.2 nm agrees with that of a monolayer, suggesting that the 1D structure corresponds to a column pulled out of the β-form crystal. A possible model of this 1D structure is schematically presented in the inset of Fig. 7.10.

Porphyrin assemblies are of fundamental importance as building blocks for the construction of functional molecular devices owing to their energy and electron-transfer functions. In particular, cyclic arrays of porphyrins

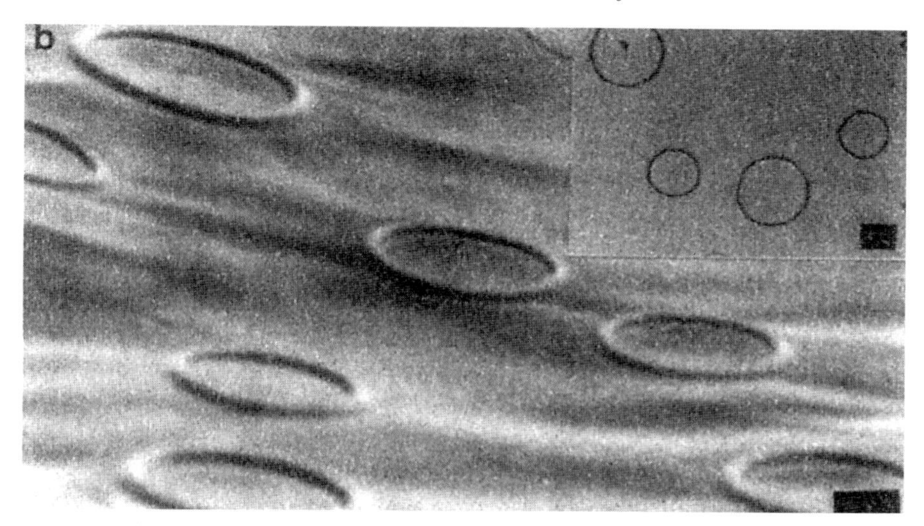

Fig. 7.11. Transmission electron microscope images of ring-shaped assemblies formed by evaporation of a thin film of porphyrin dimer. The bar represents 1 μm. The sample has been shaded with Pt (from A.P.H.J. Schenning et al., J. Am. Chem. Soc. **118**, 8549 (1996), reprinted with permission)

are attractive as architectural mimics of a light-harvesting antenna. Recently, nanometer-to-micrometer-scale ring-shaped assemblies of porphyrins formed by the generation of two-dimensional gas bubbles have been reported (Fig. 7.11). This physical phenomenon also occurs in various other systems such as nanoparticles [23] and carbon nanotubes [24], when the solvent of a solution film is allowed to evaporate on a substrate. However, the internal architecture of these rings depends strongly on the molecular species and processing. The hexakis porphyrin molecule (Fig. 7.12), which is synthesized by the coupling of six porphyrin moieties to a central benzene core via an ether linkage, leads to an increase in intermolecular interaction, resulting in a molecular ordering on a micrometer scale [25]. Hexakis porphyrin films exhibit a red shift in UV–visible spectra, implying that an intermolecular coupling arises upon aggregation. The orientation of this molecule within the ring-shaped assemblies has been revealed by a local-fluorescence investigation with polarized excitation light by near-field scanning optical microscopy (NSOM). The results indicate that the hexakis porphyrins within the ring are standing vertically on the substrate surface and their molecular planes are aligned orthogonal to the tangent of the rings, after the sample has been annealed at 80 °C for 2 days. It can be conclude that π–π stacking interactions between the molecules are important in generating these circular solid domains.

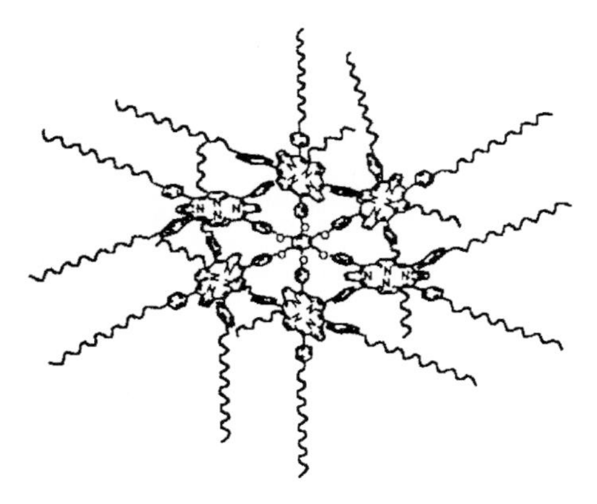

Fig. 7.12. Structure of the hexakis porphyrin molecule (from H.A.M. Biemans et al., J. Am. Chem. Soc. **120**, 11054 (1998), reprinted with permission)

7.6 Concluding Remarks

Supramolecular materials have mostly been investigated in dissolved systems, in which each molecule can move freely. For this reason, the reactions, structures and functions of supramolecular systems have been discussed only as averaged phenomena. In recent years, the development of scanning probe microscopy has enabled us to observe nanometer-scale structures of supramolecular assemblies. Although adsorbed molecules are strongly affected by molecule–surface interactions, the recent studies described in this chapter have demonstrated that supramolecular structures can be formed and controlled on surfaces. In some cases, supramolecular formation is promoted by the fact that the molecules are confined to a two-dimensional space on the substrate surface, but in other cases, it is suppressed by sticking of the molecules to specific sites on the surface.

Despite the progress in surface supramolecular chemistry, it does not yet provide a reliable route to achieving molecular devices and electronics, owing to the lack of a methodology to connect a nanoscale supramolecular system to a macroscopic electrode. For example, the current technology of microfabrication is not accessible to a single-molecular system, because the height and roughness of nanoscale electrodes are much larger than the size of the molecules. Such problems can be generalized to the subject of how to link bottom-up and top-down processes, which is one of the central issues of nanoscience and nanotechnology. Although many breakthroughs are required to realize molecular electronics, the supramolecular approach is expected to become a general strategy for the rational design and construction of desired molecular architectures with a well-defined size and shape on surfaces. We believe that this approach is a promising method for the bottom-up processing of nanoscale molecular devices.

References

1. A. Aviram, M.A. Ratner: Chem. Phys. Lett. **29**, 277 (1974)
2. A. Aviram: J. Am. Chem. Soc. **110**, 5687 (1988)
3. M.A. Reed: Proc. IEEE **87**, 652 (1999)
4. C. Joachim, J.K. Gimzewski, A. Aviram: Nature **408**, 541 (2000)
5. Y. Wada, M. Tsukada, M. Fujihira, K. Matsushige, Y. Ogawa, M. Haga, S. Tanaka: Jpn. J. Appl. Phys. **39**, 385 (2000)
6. J.M. Tour, M. Kozaki, J.M. Seminario: J. Am. Chem. Soc. **120**, 8486 (1998)
7. J.-C. Chen, M. Conrad: Physica D **75**, 417 (1994)
8. R.M. Metzger: J. Mater. Chem. **9**, 2027 (1999)
9. V.L. Langlais, R.R. Schlittler, H. Tang, A. Gourdon, C. Joachim, J.K. Gimzewski: Phys. Rev. Lett. **83**, 2809 (1999)
10. X.D. Cui, A. Primak, X. Zarate, J. Tomfohr, O.F. Sankey, A.L. Moore, T.A. Moore, D. Gust, G. Harris, S.M. Lindsay: Science **294**, 571 (2001)
11. J.-M. Lean: *Supramolecular Chemistry: Concept and Perspectives* (VCH, Weinheim 1995)
12. V.M. Hallmark, S. Chiang, J.K. Brown, C. Völl: Phys. Rev. Lett. **66**, 48 (1991)
13. M. Bohringer, K. Morgenstern, W.D. Schneider, R. Berndt, F. Mauri, A. de Vita, R. Car: Phys. Rev. Lett. **83**, 324 (1999)
14. J.V. Barth, J. Weckesser, C. Cai, P. Gunter, L. Burgi, O. Jeandupeux, K. Kern: Angew. Chem. Int. Ed. **39**, 1230 (2000)
15. T. Yokoyama, S. Yokoyama, T. Kamikado, Y. Okuno, S. Mashiko: Nature **413**, 619 (2001)
16. T. Yokoyama, S. Yokoyama, T. Kamikado, S. Mashiko: J. Chem. Phys. **115**, 3814 (2001)
17. R.H. Vreekamp, J.P.M. van Duynhoven, M. Hubert, W. Verboom, D.N. Reinhoudt: Angew. Chem., Int. Ed. **35**, 1215 (1996)
18. H.-A. Klok, K.A. Jolliffe, C.L. Schauer, J. Prins, J.P. Spatz, M. Möller, P. Timmerman, D.N. Reinhoudt: J. Am. Chem. Soc. **121**, 7154 (1999)
19. M. Furukawa, H. Tanaka, T. Kawai: Surf. Sci. **445**, 1 (2000)
20. C. Ludwig, R. Strohmaier, J. Peterson, B. Gompf, W. Eisenmenger: J. Vac. Sci. Technol. B **12**, 1963 (1994)
21. Y. Naitoh, T. Matsumoto, K. Sugiura, Y. Sakata, T. Kawai: Surf. Sci. **487**, L534 (2001)
22. A.P.H.J. Schenning, F.B.G. Benneker, H.P.M. Geurts, X.Y. Liu, R.J.M. Nolte: J. Am. Chem. Soc. **118**, 8549 (1996)
23. P.C. Ohara, J.R. Heath, W.M. Gelbart: Angew. Chem., Int. Ed. **36**, 1078 (1997)
24. J. Liu, H. Dai, J.H. Hafner, D.T. Colbert, R.E. Smalley, S.J. Tans, C. Dekker: Nature **385**, 780 (1997)
25. H.A.M. Biemans, A.E. Rowan, A. Verhoeven, P. Vanoppen, L. Latterini, J. Foekema, A.P.H.J. Schenning, E.W. Meijer, F.C. de Schryver, R.J.M. Nolte: J. Am. Chem. Soc. **120**, 11054 (1998)

8 Semiconductor and Molecular-Assembly Nanowires

Tomoyuki Akutagawa, Takayoshi Nakamura

Summary. Highly conducting nanowires are expected to be critical for functionalizing and integrating nanoscale electronic devices. Semiconductor nanowires are more important than metal nanowires from the viewpoint of device applications. Recent research on the preparation and fabrication of inorganic semiconductor nanowires were briefly reviewed. Molecular-assembly nanowires will have an important role in the complete bottom-up manufacture of molecular electronics, whose devices are built up from synthesized molecules through self-assembly processes. Such nanowires can be assembled from π-molecules through molecule-by-molecule π-stacking. Research on molecular conductors will offer guiding principles for constructing molecular nanowires with appropriate electronic properties. At the same time, supramolecular chemistry will offer powerful methods to build up molecular nanowires through self-assembly processes. Three kinds of molecular nanowires, tetrathiafulvalene (TTF)-halide, crown-ether-fused phthalocyanine, and amphiphilic TTF macrocycle are introduced as molecular-assembly nanowires composed of molecular conductors.

8.1 Introduction

Serious difficulties in traditional semiconductor manufacturing techniques (the top-down method) will be reached in the next decade [1,2]. One potential long-term solution to the obstacles to computer development that will appear to an increasing extent is to shift the basic operations of computation to devices that take advantage of molecular electronics [2]. The bottom-up approach, in which objects are assembled molecule-by-molecule (self-assembly), could be used to construct a chemical computer from a large number of chemically assembled logic switches [3,4]. The bottom-up chemical approach has the potential to realize a nanoscale electronic device, which is built up from synthesized molecules through self-assembly processes. One of the advantages of this approach is a significantly lower power consumption during the manufacturing of individual devices. However, several important outstanding issues remain in the self-assembly chemical approach [2–4]. For example, a molecular device will need to hold a memory state for a long period, just as in present computers. However, the chemical stability of ordinary molecules is much lower than that of inorganic semiconductors. Also, the manufacturing and self-assembly processes used for these molecular-scale devices will lead to a large number of defects in the device structure. A defect-tolerant computer architecture will be needed, and a prototype small-scale system has been proposed by Heath et al. [4]. The progress to date in molecular electronics has been substantial, and thus it seems possible that a self-assembly molecular computer will be developed over the next 10 years that would be relatively attractive compared with conventional computers [2–4].

Nanoscale electronic systems using inorganic semiconductors, containing zero-dimensional (0D) quantum dots, one-dimensional (1D) wires, and two-dimensional (2D) electron gases, have been prepared by top-down manufacturing techniques, and the transport properties of these systems extensively studied [5]. The fabrication of these 0D, 1D, and 2D fundamental nano structures in molecular system through a bottom-up approach will be one of the most important steps in realizing nanoscale electronic devices. The 1D systems are the structures with the lowest dimension that permit efficient electron transport. We often see the importance of 1D nanostructures in biological systems [6,7]. There are a variety of 1D nanostructures here: for example, the double helix of DNA, with a diameter of $\sim 2\,\mathrm{nm}$, the α-helix structure of proteins, with a diameter of $\sim 0.5\,\mathrm{nm}$, and the axons in the neural systems, with a size $\sim 10\,\mathrm{nm}$, are incorporated into the hierarchal structure of a living body. Nanowires are expected to be critical for functionalizing and integrating nanoscale electronic devices. Highly conducting nanowires will be important units in constructing electronic circuits, and thus semiconductor, metallic, and superconducting nanowires are being extensively studied with the aim of developing nanoscale electronic system.

A wide range of compounds, including inorganic metals (Au, Pt, Ni, etc.), semiconductors (Si, GaN, InP, etc.), carbon nanotubes, polymers (polyacetylene, poly-*p*-vinylene, polypyrrole, etc.), molecular conductors, π-conjugated molecules (terphenyl, oligo(thiophene), oligo(phenylene ethylene), etc.) and DNA have been used as nanowires [3]. The diameter, length, and electronic structure of these nanowires vary significantly. Figure 8.1 depicts several nanowires together with their electronic structure. The electronic properties of nanowires range from insulating, semiconducting, and metallic to superconducting. We have classified these nanowires into two categories on the basis of their electronic structure for the sake of convenience in this text. The first category contains those whose electrical conduction is dominated by the carriers around the Fermi level of the band structure, just as in bulk metals and semiconductors. Although the electronic density of states along the width of a nanowire has a discrete character because of the restricted lattice translation along this axis, it can be approximately represented as a band structure along the wire direction for a typical length of several micrometers. Doped polymers, single-walled carbon nanotubes (SWCNTs), and semiconductor and metal nanowires may have Fermi surfaces. On the other hand, DNA is an insulator with a large band gap, although its electrical properties are still controversial. The second category contains low-molecular-weight π-conjugated

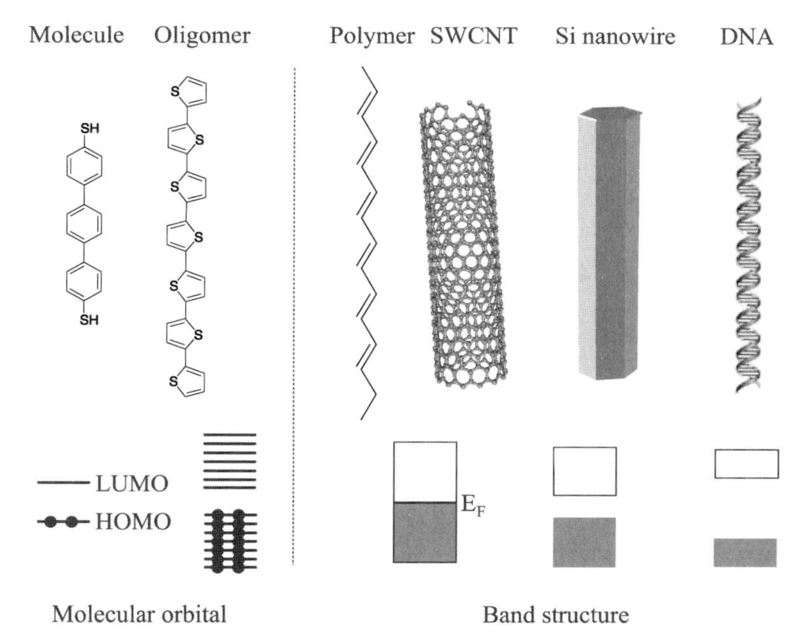

Fig. 8.1. Nanowires constructed from single molecules, oligomers, polymers, single-walled carbon nanotubes (SWCNTs), semiconductors, and DNA. The electronic structures of these are illustrated at the bottom of the figure. Low-molecular weight π-conjugated molecules have discrete molecular orbitals, while the doped polymers, SWCNTs, semiconductor nanowires, and DNA have band structures

molecules such as terphenyl, the electronic state of which can be represented as discrete molecular orbitals (HOMO and LUMO) far from a band structure [3,8]. It has been suggested theoretically that the electrical conduction mechanism of a single molecule is dominated by resonant tunneling between the HOMO (or LUMO) and the Fermi levels of metal electrodes. A energy mismatch between a molecular orbital and the Fermi level of a metallic electrode simply causes a flow of a tunneling current between the metallic electrodes through an insulating single molecule. Recently, an extremely large π-conjugated oligomer of *meso-meso*-linked zinc(II) oligoporphyrins having 62 porphyrin units has been prepared [9]. The electronic structure of this kind of molecule is expected to exist in the intermediate region between above two categories. Single-molecule nanowires, whose electronic structure is dictated by molecular orbitals, are beyond the scope of this chapter. We have focused on the first category of nanowires, whose electronic structure can be described by a band structure.

Remarkable progress in the field of semiconductor nanowires and SWCNTs, directed towards the realization of nanoscale computers, has been witnessed in the last decade. Semiconductor nanowires are more important than metal nanowires from the viewpoint of device applications. Recent research on semiconductor nanowires is briefly reviewed in Sect. 8.2. Owing to limitations of space, we shall not go into SWCNT devices, and readers might turn to recent references [10]. Although only a few examples of the preparation of nanowires of molecular assemblies have been reported, we consider that these molecular nanowires will have an important role in the complete bottom-up manufacture of molecular electronics. Such nanowires can be assembled from π-molecules through molecule-by-molecule π-stacking. Research in the field of molecular conductors will offer guiding principles for constructing molecular nanowires with appropriate electronic properties. The anisotropic charge transfer interaction in molecular conductors is advantageous for forming a 1D π–π stacking nanowire structure. At the same time, supramolecular chemistry will offer powerful methods to build up molecular nanowires through self-assembly processes. We introduce three kinds of molecular nanowires composed of molecular conductors in Sect. 8.3.

8.2 Semiconductor Nanowires

A large number of semiconductor nanowires, such as Si, GaN, and InP nano wires, have been prepared so far [11–14]. The electrical-conduction properties of semiconductor nanowires and the fabrication of devices from them have been extensively examined; in particular, preparation methods to control both the size and the electronic state have progressed remarkably [15,16]. Nanowires with diameters in the range from several tens to hundreds of nanometers and lengths of several micrometers have been prepared. Both p- and n-type semiconducting nanowires have been obtained by controlling the doping level [17,18]. A nanoscale $p-n$ junction has also been realized by the combination of p- and n-type semiconducting nanowires; this allows the creation of device prototypes such as a diode, a field effect transistor (FET), and a bipolar transistor [19,20]. Multiple connection of these elements gives AND, OR, and NOR logic gates,which could be used in nanoscale computation [20]. Furthermore, nanoscale biological and chemical sensors [21], as well as photoluminescence and photodetection devices [22], utilizing semiconductor nanowires, have been proposed.

In these devices, the semiconductor nanowires have a variety of electronic structures and act as central elements of the device rather than as electrical leads connecting active units. One of the typical devices in which semiconductor nanowires play a fundamental role is the crossbar switch. This contains a nanoscale point contact between crossing nanowires; this region acts as an active device element, which may be $p-n$ junction or an FET unit (Fig. 8.2a) [19,20]. It has been suggested that a crossbar switch array may become a basic architecture for constructing defect-tolerant computing systems [4]. A nanowire device based on a crossbar arrangement of SWCNTs was first proposed by Rueckes et al. [23]. These authors pro-

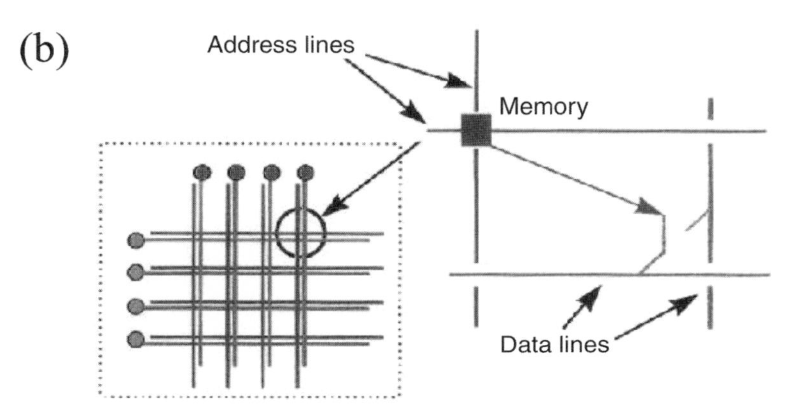

Fig. 8.2. (**a**) Schematic illustration of the crossbar switch. The nanoscale crossing points act as active device elements of a nanoscale electronic device (from T. Rueckes et al., Science **289**, 94 (2000), reprinted with permission). (**b**) A concept of a crossbar switch device used in a computing system in which each crossing point acts as a memory element (from J.R. Heath et al., Science **280**, 1716 (1998), reprinted with permission)

posed applying a square lattice of SWCNTs to create bistable electrostatic ON/OFF switches, for constructing nonvolatile random access memory. This idea was expanded to a crossbar semiconductor nanowire structure, as shown in Fig. 8.2.

8.2.1 Preparation and Assembly

Several methods, such as chemical vapor–liquid–solid (VLS) reaction [24], template synthesis [25–28], solution growth [29,30], electrodeposition [31], and laser-assisted catalytic growth [11,32] have been used for the preparation of semiconductor nanowires. A critical reaction process is the promotion of anisotropic crystal growth from a nanoscale seed particle. The VLS reaction is a typical anisotropic crystal growth method in the gas phase [24]. Laser-assisted catalytic growth is also utilized in the form of a modified VLS reaction, by which method precise control over both the diameter and the electronic state of the nanowires becomes possible. The diameter of the semiconductor nanowires is dominated by the size of the seed particles used in the VLS reaction.

Figure 8.3 shows a schematic view of template synthesis (Fig. 8.3a,b), solution growth (Fig. 8.3c), and laser-assisted catalytic growth (Fig. 8.3d). Template synthesis is a powerful method for nanowire growth. In the template growth of nanowires, ordered nanoporous structures are utilized. For example, in the electrodeposition of semiconducting materials in nanoporous structures, track-etched membranes, or anodized aluminum, nanowires grow in the nanoscale 1D pores [25]. Both the diameter and the length of the nanowires are regulated by the size of the nanopores.

High-density, oriented metal nanowires (Co, Ni, and Ag) have been prepared by using self-assembled polymers or organic nanotubes as templates [26,27]. Template nanowire synthesis using a polymer matrix was reported first for $(Mo_3Se_3^-)_\infty$, where the wire diameter ranged from 0.6 to 2 nm and the length was about 100 nm [33]. A rapid polymerization of vinylene carbonate in the presence of a cross-linking agent gave a random-coil structure in which the $(Mo_3Se_3^-)_\infty$ nanowires grew. Since the template space in the polymer matrix was randomly distributed, both the density and the orientations of the $(Mo_3Se_3^-)_\infty$ nanowires were difficult to control during this process. To improve the situation, phase separation phenomena of block copolymers were utilized to regulate the template structure [26]. A phase-separated, cylindrical, hexagonal lattice of polymethylmethacrylate (PMMA) in a polystyrene (PS) matrix was formed in a $1:1$ block copolymer on an Au substrate (Fig. 8.3a). The cylindrical phase of PMMA, with a diameter of 14 nm, was oriented uniaxially by the application of a dc electric field. After selective removal of the PMMA by UV irradiation, copper or cobalt was introduced by an electrodeposition procedure in the vacant nanopores in the PS matrix. An ultrahigh-density nanowire array with a nanowire density of 1.9×10^{11} nanowires/cm^2 was prepared by this phase-separated block-copolymer technique.

A high-density nanowire array of Ag single crystals has been prepared by a solution growth method using self-assembled calix[4]hydroquinone nanotubes through a photochemical redox reaction in solution (Fig. 8.3b) [27]. Calix[4]hydroquinone forms a nanotube assembly through hydrogen bonding in aqueous solution. Crystal structural analysis revealed the formation of nanoscale pores with a diameter of ~ 0.6 nm and a nearest-neighbor interpore distance of 1.7 nm. In crystals of this material, the pores are filled with water molecules, which can be replaced by Ag^+ ions in acidic conditions. Since the surface of the nanopores is composed of π-planes of the hydroquinone moiety, the Ag^+ ions can easily penetrate the nanopores through a cation–π interaction. Photochemical reduction of Ag^+ ions in the nanopores gave single-crystal Ag nanowires with a diameter of 0.4 nm. The metallic Ag nanowires

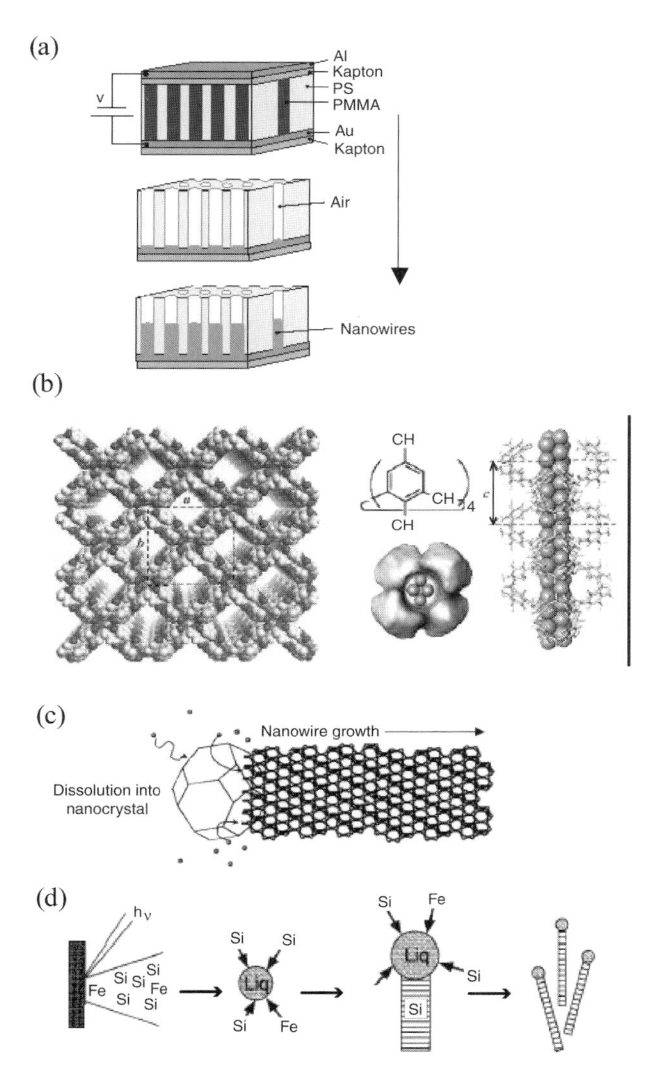

Fig. 8.3. Preparation methods for nanowires. (**a**) Template growth of metal nanowires in phase-separated block copolymer. A phase-separated cylindrical hexagonal lattice of PMMA in the PS matrix acts as a template (from T. Thurn-Albrecht et al., Science **290**, 2126 (2000), reprinted with permission). (**b**) Array of Ag single-crystal nanowires in templates of self-assembled calix[4]hydroquinone nanotubes (from B.H. Hong et al., Science **294**, 348 (2001), reprinted with permission). (**c**) Diameter-selective synthesis of Si nanowires in supercritical fluid solution. Au nanoparticles coated with alkanethiol are the seed particles (from J.D. Holmes et al., Science **287**, 1471 (2000), reprinted with permission). (**d**) Laser–ablation method for the growth of Si nanowires (from A.M. Morales et al., Science **279**, 208 (2000), reprinted with permission)

are surrounded by the organic host and are quite stable in solution or even in air.

Nanowire preparation in solution is a convenient and flexible method. Without templates, isotropic crystal growth is usually observed in the solution phase. A novel method to control anisotropic crystal growth and the size of semiconductor nanowires without any template has recently been developed in supercritical fluid solution [30]. To obtain anisotropic crystal growth, Au nanoparticles coated with alkanethiol with an average diameter of 2.5 nm were employed as seed particles (Fig. 8.3c). The precursor diphenylsilane decomposed to atomic Si in supercritical hexane (500 °C and 270 bar), which dissolved into the Au nanoparticles forming an Si–Au binary alloy. The supersaturated atomic Si in the Si–Au alloy crystallized along the [111] direction to form nanowires. The single-crystal Si nanowires formed by this process had an almost uniform diameter, distributed around 4 to 5 nm. The formation of Si nanowires was observed only under reaction conditions above the critical point.

The laser-ablation method is widely employed in the preparation of semiconductor nanowires, such as those of doped Si, binary III–V compounds (GaAs, GaP, InAs, and InP), binary II–VI compounds (ZnS, ZnSe, CdS, and CdSe), and ternary compounds (GaAs/P and InAs/P) [11,12]. The crystallization mechanism in this method is basically similar to that in the VLS method [25]. Figure 8.4 shows transmission electron microscope images of Si nanowires prepared by the laser-ablation method. Liquid alloys such as Si–Au, Fe–Au, GaAs–Au, and InP–Au, produced by the laser-ablation process, are used as catalytic seed particles for the anisotropic crystal growth. For example, Si nanowires have been prepared by the laser ablation with a Nd-YAG laser at a wavelength of 532 nm of a binary $Si_{0.9}Fe_{0.1}$ target at 1200 °C in a reaction chamber. Here, vapors of atomic Si and Fe generated by laser ablation from the target are condensed into liquid Fe–Si nanoclusters through cooling by collision processes with the buffer gas (Fig. 8.3d). Owing to the continuous supply of atomic Si to the liquid Si–Fe nanoclusters, the Si atoms are supersaturated in the liquid nanoparticles, and crystallize into single-crystalline Si nanowires. The yield of semiconductor nanowires obtained by this method is over 90%, and the Si nanowires grow along the [111] direction. However, the liquid nanoparticles formed by the laser-ablation procedure have a diameter distribution, which influences the diameters of the nanowires in a reaction batch. The diameter distribution was improved when size-defined Au nanoparticles deposited on a Si substrate in advance were used [17].

The electronic properties of semiconductor nanowires have been precisely controlled by impurity doping. Both *p*- and *n*-type impurity doping of Si nanowires has been successfully carried out during the laser-ablation process. *p*- and *n*-type dopings are achieved by introducing boron (B) and phosphorus (P), respectively. *p*-type Si nanowires are formed by the laser ablation of an Au target in the presence of reaction gases containing of SiH_4 and B_2H_6, while

n-type Si nanowires are obtained by using an Au–P target and red phosphorus. The doping level can be controlled by changing the ratio of the reaction gases from a value that provides light doping ($SiH_4 : B_2H_6 = 1000 : 1$) to one that provides heavy doping ($SiH_4 : B_2H_6 = 2 : 1$). The control of the electronic state of semiconductor nanowires is one of the key technologies for constructing practical nanoscale electronic devices.

The nanowires obtained from the reaction chamber are entangled with each other, as shown in Fig. 8.4. It is necessary to disentangle them and arrange them on a substrate for electrical measurements or device fabrication. It should be noted that the surfaces of disentangled single Si nanowires are covered with an oxide layer (Fig. 8.4). A simple and convenient method to disentangle the nanowires is solution casting of a highly diluted dispersion of nanowires. This procedure will arrange nanowires randomly on the substrate surface. Recently, several approaches have been attempted to align nanowires on a substrate along a specified direction. One example is the fluidic alignment of semiconductor nanowires on a patterned surface [34,35]. Other approaches that have been used are the electric-field-assisted assembly of metallic Au nanowires between patterned electrodes [36], an ion-beam-directed self-assembly technique for metallic calcium nanowires [37], self-aggregation of chemically modified Ag nanocrystals at an air–water interface [38], and dielectrophoretic assembly of nanowires from a nanoparticle suspension [39]. We describe the fluidic alignment technique here.

The flow of liquid in a ditch structure can control the flow direction of nanowires. In the fluidic alignment technique, a dispersion of nanowires in CH_3OH is introduced into a one-dimensional channel structure on a substrate

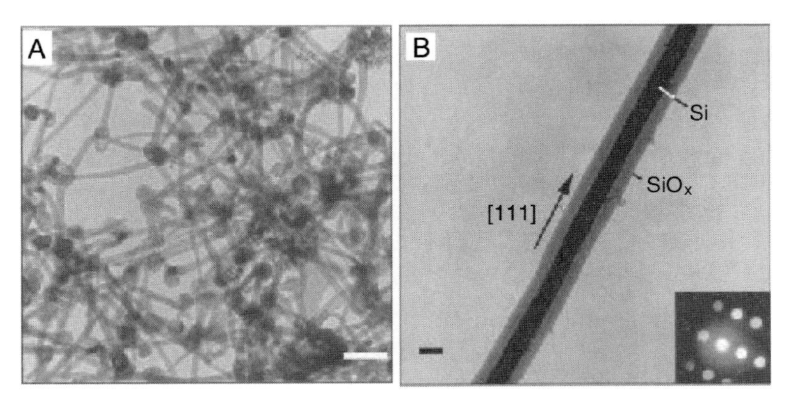

Fig. 8.4. Transmission electron microscope images of Si nanowires prepared by the laser-ablation method. The *white* and *black scale bars* in the *left* and *right figures* correspond to 100 nm and 10 nm, respectively. The single-crystal Si nanowires grow along the [111] direction, and the surface of the Si nanowire is covered with an oxide SiO_x layer (from A.M. Morales et al., Science **279**, 208 (2000), reprinted with permission)

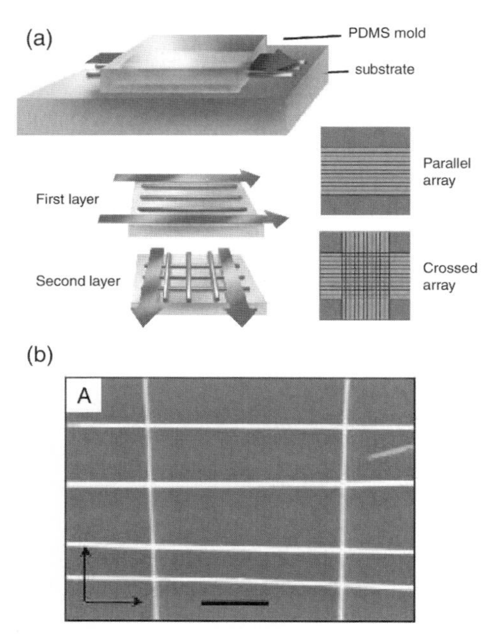

Fig. 8.5. Fluidic alignment of nanowires on a substrate surface. (**a**) PDMS mold on substrate. A dispersion of nanowires is introduced into the PDMS mold. A crossed array is formed by layer-by-layer fluidic alignment (from R.F. Service, Science **293**, 782 (2001), reprinted with permission). (**b**) SEM image of InP crossbar structure formed by a two-step layer-by-layer fluidic assembly method. The scale bar corresponds to 500 nm layer (from Y. Huang et al., Science **291**, 630 (2001), reprinted with permission)

formed by using a poly(dimethylsiloxane) (PDMS) mold (Fig. 8.5) [35]. The nanowires become aligned along the channels with the aid of fluidic flow. The nanowires can be arranged on a millimeter-scale SiO_2 substrate with an average internanowire separation of ~ 100 nm. By a stepwise procedure in which the flow direction is changed, a crossed array of semiconductor nanowires can be formed (Fig. 8.5b). Chemical modification of the SiO_2 substrate surface can be used to control the internanowire separation obtained from the fluidic alignment technique. A hydrophobic–hydrophilic stripe pattern is formed on the SiO_2 surface from self-assembled monolayers of hexamethyldisilazane (HMDS) and 3-aminopropyltriethoxysilane (APTES). Upon the fluid flow of semiconductor nanowires over this patterned substrate surface, the nanowires are selectively assembled on the NH_2-terminated surface regions.

8.2.2 Electrical Properties

The electrical transport properties of undoped, *p*-type, and *n*-type Si nanowires have been evaluated using a two-probe electrode with a gap of several micrometers [17,18,41]. The gate voltage dependence of the I–V characteristic has been also measured in these nanowires. Since the surfaces of Si nanowires are oxidized to an amorphous SiO_x layer, as shown in Fig. 8.4, a nonohmic I–V characteristic is observed in undoped Si nanowires. Figure 8.6a shows the gate voltage dependence of the I–V characteristic of an undoped Si nanowire with a diamter of 70 nm. A negative gate bias enhanced the source–drain current, indicating p-type behavior of the undoped Si nanowire. The electrical resistivity for zero gate voltage was $3.9 \times 10^2\ \Omega\,\mathrm{cm}$ at room temperature.

The lightly doped Si nanowire was ohmic with a resistivity of $1\ \Omega\,\mathrm{cm}$, which is two orders of magnitude lower than that of the undoped Si nanowire. The gate voltage dependence of the I–V characteristic is as expected for a

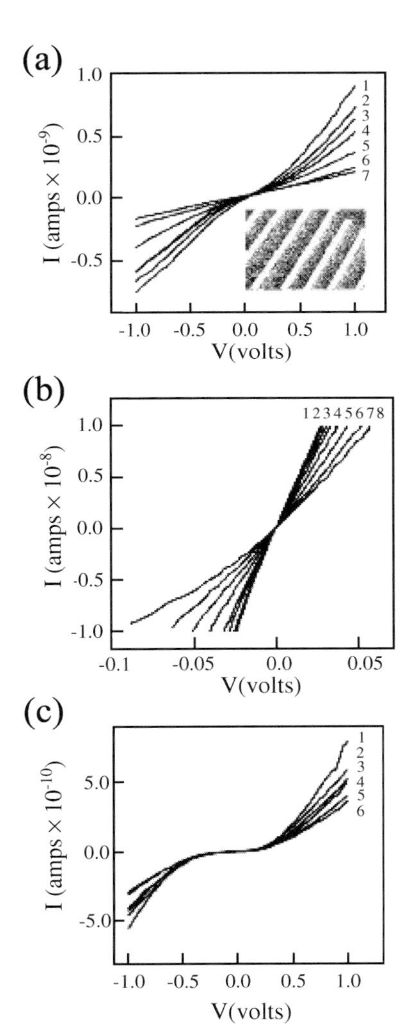

Fig. 8.6. Gate voltage dependence of the I–V characteristics of (**a**) an undoped Si nanowire, (**b**) a lightly doped p-type Si nanowire, and (**c**) a lightly doped n-type Si nanowire (from Y. Cui et al., J. Phys. Chem. B **104**, 5213 (2000), reprinted with permission)

p-type semiconductor (Fig. 8.6b). On the other hand, a heavily B-doped Si nanowire had a low electrical resistivity of $6.9 \times 10^{-3}\,\Omega\,\mathrm{cm}$ and its linear *I*–*V* characteristic was independent of the gate voltage. This result suggests a semimetallic or metallic electronic state for heavily B-doped Si nanowires.

The *n*-type Si nanowires used here were obtained by phosphorus doping. A nonlinear *I*–*V* characteristic was observed for a lightly *n*-doped nanowire owing to the nonohmic electronic contacts with the electrodes (Fig. 8.6c). The electrical resistivity, $2.6 \times 10^{2}\,\Omega\mathrm{cm}$, is of the same order as that of the undoped Si nanowire. A heavily *n*-doped Si nanowire showed a quite significantly lower electrical resistivity of $2.3 \times 10^{-2}\,\Omega\,\mathrm{cm}$. A gate voltage dependence was not observed, which suggests a semimetallic or metallic electronic state, as in the case of the heavily doped *p*-type Si nanowire. The electronic mobilities of the undoped and lightly doped *n*-type Si nanowires were estimated to be 5.9×10^{-3} and $3.17\,\mathrm{cm}^2/(\mathrm{Vs})$, respectively. The mobility was slightly lower than that of bulk Si ($\sim 10\,\mathrm{cm}^2/(\mathrm{Vs})$) [40]; this differnce was attributed to the effect of the nanoscale size. The small diameter (60 – 70 nm) of the nanowires enhances the probability of electronic scattering along the width of the nanowire.

8.2.3 *p*–*n* Junctions and FET Devices

The use of a stepwise fluidic technique for nanowire alignment on an SiO_2 substrate surface make it possible to arrange *p*- and *n*-type nanowires into a crossbar structure [19,35]. The point contacts between the bars where they act as nanoscale *p*–*n* junctions. Figure 8.7a shows a field-emission scanning electron microscopy (FESEM) image of a crossbar structure of *p*- and *n*-type Si nanowires. The transport properties of *n*–*n*, *p*–*p*, and *p*–*n* crossbar junctions of lightly doped Si nanowires (20 – 50 nm) have been examined. Linear and nonlinear *I*–*V* characteristic were observed for the *p*–*p* and *n*–*n* crossbar structures, respectively. The thick surface oxidation layers on the *n*-type Si nanowires act as insulating layers between the nanowires or between the nanowires and the electrodes, which was the origin of the nonlinear *I*–*V* characteristic.

A *p*–*n* crossbar junction structure of *p*- and *n*-type Si nanowires showed a rectification effect (Fig. 8.7). However, the thick surface oxide layer of *n*-type Si nanowires makes it difficult to control the electronic properties of crossbar *p*–*n* junctions. For this reason, stable *n*-type GaN nanowires were employed in a crossbar arrangement instead of Si nanowires [20]. The *n*-type GaN nanowires showed ohmic contacts to the electrodes and were quite stable under ambient conditions. A multipoint crossbar arrangement was constructed from a combination of *n*-type GaN and *p*-type Si nanowires using a layer-by-layer fluidic flow technique (Fig. 8.7b). The rectification effect of the *p*–*n* crossbar junction was improved in this structure, and the turn-on voltage and rectification ratio were around 1 V and 95%, respectively (Fig. 8.7b) [20].

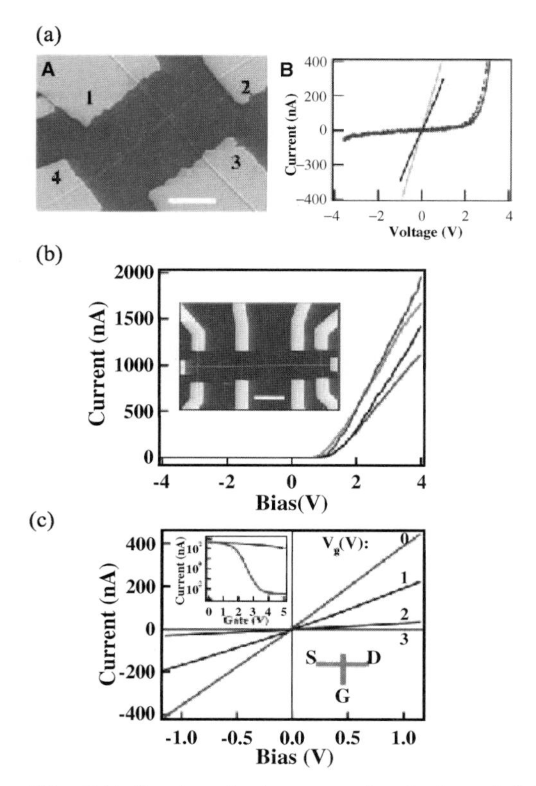

Fig. 8.7. Semiconductor nanowire devices. (**a**) I–V characteristic (*right*) and field-emission scanning electron microscopy (FESEM) image (*left*) of a crossbar structure of p- and n-type Si nanowires. The *white scale bar* corresponds to 2 μm. The nanowires are connected to four electrodes, 1–4; a p-type nanowire is connected to electrodes 1 and 3, and an n-type nanowire to 2 and 4. The I–V characterstic measured between electrodes 1 and 2 shows a rectification effect (from Y. Cui et al., Science **291**, 851 (2001), reprinted with permission). (**b**) SEM image and I–V characterstics of a nanowire array device constructed from four p-type Si and one n-type GaN nanowire connected to ten electrodes. The characterstics show a turn-on voltage at around 1 V. (**c**) Gate-dependent (0, 1, 2, and 3 V) I–V characterstics of a crossbar p–n junction structure formed by p-type Si and n-type GaN nanowires. The current flow in the p-type Si nanowire was controlled by the application of a gate voltage to the GaN nanowire (from Y. Huang et al., Science **294**, 1313 (2001), reprinted with permission)

A crossbar structure between p-type Si and n-type GaN nanowires can also act as a field-effect transistor (FET) (Fig. 8.7c). The extremely thin SiO_x oxide layer ($\sim$ 1 nm) on the surface of the p-type Si nanowire behaves as a nanoscale gate insulator. The current flow in the p-type Si nanowire was controlled by the application of a gate voltage to the GaN nanowire. An increase in the gate voltage on the n-type GaN electrode toward the positive direction decreased the current in the p-type Si nanowire. This nanoscale FET device has a high performance, with an ON/OFF ratio of 10^5 under the application of a gate voltage of about 1 V. A device prototypes of logic gates (OR, AND, and NOR) was prepared by combining nanowire-based FET structures. Further integration of these logic gates will realize a nanoscale computing system based on semiconductor nanowires.

8.3 Nanowires of Molecular Assemblies

In contrast to the degree of success that has been achieved with semiconductor-based nanowire devices, the research on the application of molecular assemblies to active device structures has just started. However, because of the flexibility in molecular design together with the rich physics of molecular solids, molecular-assembly nanowires are considered as quite promising candidates for constructing future nanoscale devices. Molecular nanowires are constructed using weak intermolecular interactions such as van der Waals (dipole–dipole, dipole–induced dipole, and dispersion interactions), charge transfer, and hydrogen-bonding interactions [41]. The crystallization of molecular materials is usually carried out from the solution phase because of the low thermal stability of organic compounds. When the molecules are relatively stable and are readily sublimed, for example aromatic hydrocarbons and phthalocyanines, gas phase crystallization can be applied.

Ordinary molecules are not stable enough against the electron-beam lithography and etching procedures typically employed for semiconductor nanowire device preparation. Therefore novel methods to prepare, arrange, and integrate nanowires constructed from molecules into devices are required. To obtain one-dimensional nanowire structures, an appropriate design of molecular shape, intermolecular interaction, and electronic structure is necessary. These nanowires are self-assembled in a bottom-up way to produce nanoscale electroactive wires. The final goal should be a complete replacement of the top-down approach of current device fabrication processes by a self-assembly bottom-up approach.

The term "supramolecular chemistry" labels one of the key sciences required to realize the bottom-up self-assembly approach [42]. A supramolecule is defined as an entity composed of several or a large number of molecules, which are connected to each other through noncovalent weak intermolecular interactions. The appropriate design of these intermolecular interactions is essential to obtaining self-assembly molecular nanowires, which are also supramolecular entities. The concept of a "programmed molecule", which contains chemical information concerning the sites, strength, directions, etc, of the intermolecular interactions, has been proposed by Lehn [42]. Self-assembled supramolecules with the desired structure can be achieved by using appropriately designed programmed molecules. An example of such a process

(a)

Hydrogen-bonding chain

Hydrogen-bonding macrocycle

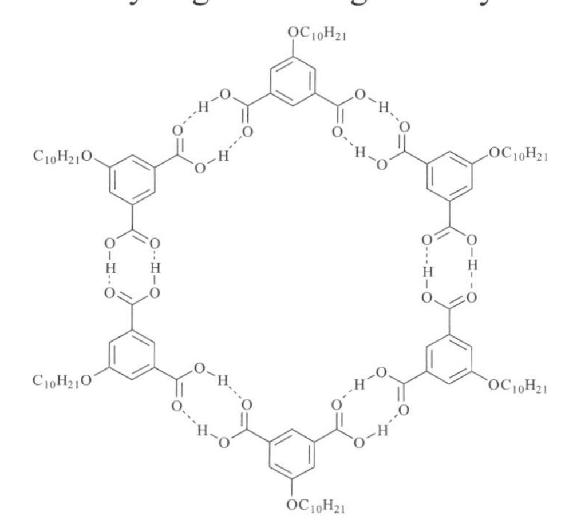

(b)

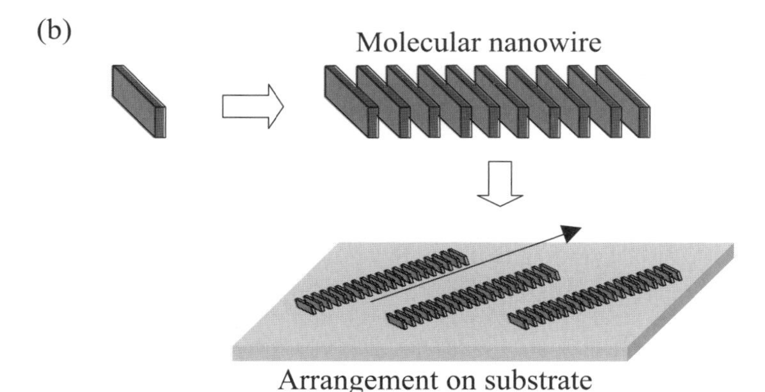

Fig. 8.8. (**a**) Self-assembled supramolecules formed through designed hydrogen-bonding sites. Phenyl rings with hydrogen-bonding sites provide self-assembled 1D chain and cyclic structures. (**b**) Formation of a molecular nanowire through the π–π stacking interaction of molecular conductors

is the formation of an infinite 1D chain or a cyclic structure through hydrogen-bonding interactions (Fig. 8.8a) [42]. Phenyl rings with hydrogen-bonding sites at the 1 and 4 positions provide self-assembled 1D chain structures, and hydrogen-bonding sites at the 1 and 3 positions provide cyclic structures.

For the design of electroactive molecular nanowire structures, we need to introduce electrical conduction (Fig. 8.8b). The introduction of a charge transfer interaction, which is also a weak intermolecular interaction [43], is a straightforward method to realize electrical conduction in nanowires. Also, it is desirable that orientation of the nanowires on the substrate surface is realized during the self-assembly process. In Sect. 8.3.1, we describe molecular conductors briefly and explain the reason why molecular conductors are suitable for the formation of molecular nanowires. There are a few examples of electroactive molecular-assembly nanowires that have been studied, three examples of which are reviewed in the subsequent sections.

8.3.1 Molecular Conductors

A large number of molecular conductors, ranging from semiconductor and metal to superconductor, have been prepared [44,45]. In general, a stable organic molecule has a closed-shell electronic structure without conduction carriers, and therefore molecular solids are highly insulating. Conduction carriers can be generated by an intermolecular charge transfer (CT) interaction between the HOMO of an electron donor (D) molecule and the LUMO of an electron acceptor (A) molecule. Figure 8.9a shows the molecular structures of some typical D and A molecules utilized in molecular conductors. Among these molecules, tetrathiafulvalene (TTF) and 7,7,8,8-tetracyano-*p*-quinodimethane (TCNQ) are well-known D and A molecules, respectively;

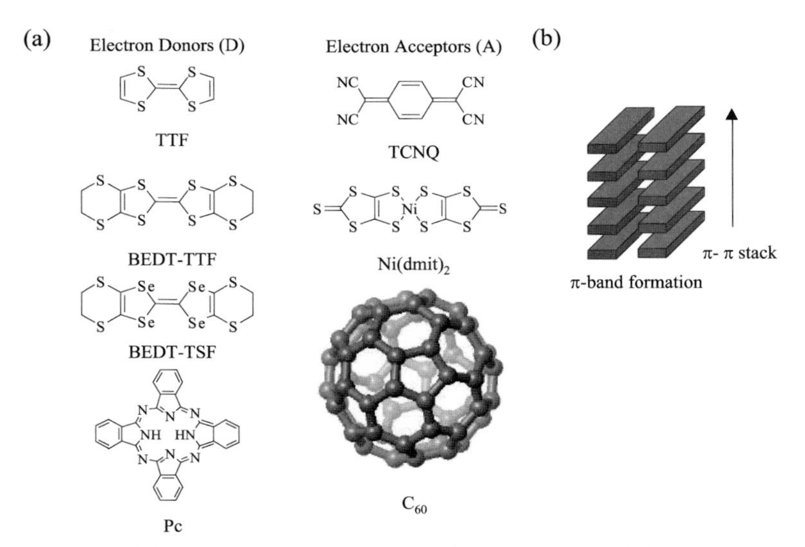

Fig. 8.9. (**a**) Typical electron donor (D) and acceptor (A) molecules employed in molecular conductors. (**b**) A 1D columnar structure produced through the π–π stacking interaction in a crystal

they gave the first molecular metal, $(TTF^{+0.59})(TCNQ^{-0.59})$. The degree of CT ($\delta$) in the binary $(D^{+\delta})(A^{-\delta})_x$ CT complex depends on the ionization potential of the D molecule and the electron affinity of the A molecule. Planar π-conjugated D and A molecules have a tendency to form a 1D columnar structure through the π–π stacking interaction (Fig. 8.9b). Electrical conduction is caused by the electrons and/or holes around the Fermi level of the π-band structure, which is formed by the stack of planar π-molecules with an open-shell electronic structure. Since the π-orbitals are orthogonal to the molecular plane, the direction of the π–π interaction is highly anisotropic. This 1D π–π interaction in molecular conductors is suitable for obtaining molecular nanowires. The examples of molecular-assembly nanowires described here are composed of stacks of planar π-molecules.

8.3.2 TTF–Br Nanocrystals

Single crystals of molecular conductors are typically millimeters in size, and needle, plate, block, or flake in shape. The shape of a single crystal is related to the dimensionality of the electronic state, and molecular conductors with 1D electronic structures have a strong tendency to form needle-shape single crystals. Downscaling of these needle-like single crystals to the nanoscale will provide molecular nanowires. Favier et al. modified an electrocrystallization method to control the crystal growth of the cation radical salt TTF–Br [46].

In the work of Favier et al., Pt nanoparticles deposited on an HOPG surface were employed as a cathode for electrocrystallization of $(TTF)(Br^-)_{0.76}$. The crystal structure of $(TTF)(Br^-)_{0.76}$ has been already solved and is composed of a regular TTF stack; the material exhibits high electrical conductivity ($\sigma = 400\,Scm^{-1}$) at room temperature [47]. Pt nanoparticles were prepared by slow electrodeposition; these were randomly distributed on the HOPG surface (Fig. 8.10a, left). The diameter of the nanoparticles could be controlled between 70 and 1300 nm by changing the electrodeposition conditions. Crystal growth of $(TTF)(Br^-)_{0.76}$ from the Pt nanoparticles was observed (Fig. 8.10a, right). Electrocrystallization was achieved by applying 1- – 15- voltage pulses to the HOPG with a period of 0.2 second, and crystal growth was observed only from the Pt nanoparticles. Figure 8.10b shows SEM images of nanowires prepared from Pt nanoparticles of different sizes. The diameter of the nanowires was proportional to that of the Pt nanoparticles, and was about 40–50% of that of the nanoparticle. The diameter of the nanowires can be varied from 25 to 600 nm and the typical length was around 1 μm. The electrical conductivity of these nanowires has not been measured yet. High electrical conductivity is, however, expected, judging from the crystal structure. This novel electrocrystallization method has the potential to form a variety of highly conducting molecular nanowires from TTF-based molecular materials.

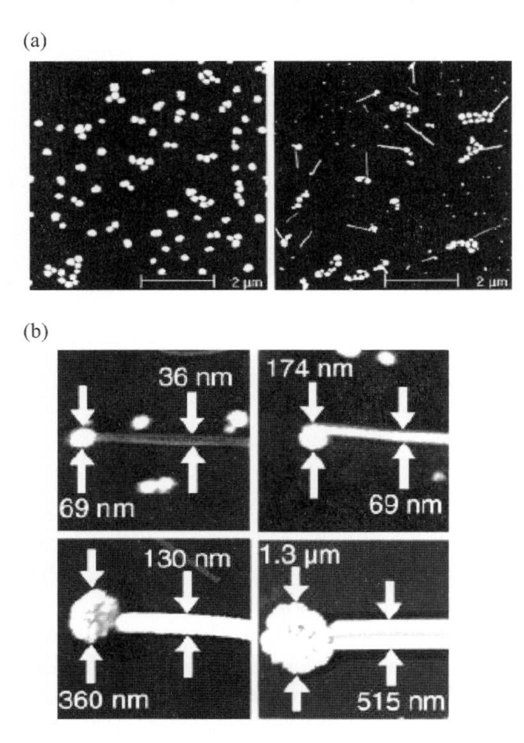

Fig. 8.10. (TTF)(Br^-)$_{0.78}$ nanocrystals prepared by an electrocrystallization method. (**a**) SEM images of Pt nanoparticles randomly distributed on an HOPG surface (*left*) and (TTF)(Br^-)$_{0.78}$ nanowires prepared by electrocrystallization (*right*). The *scale bar* corresponds to 2 µm. (**b**) SEM images of (TTF)(Br^-)$_{0.78}$ nanowires grown from Pt nanoparticles (from F. Favier et al., Adv. Mater. **13**, 1567 (2001), reprinted with permission)

8.3.3 Crown-Ether-Fused Phthalocyanine

Amphiphilic molecules form interesting molecular-assembly structures such as black lipid membranes, liposomes, vesicles, and micelles in solution. These molecular assemblies have been employed in the formation of molecular nano wires [48]. A 1D supramolecular assembly of an amphiphilic crown-ether-fused phthalocyanine (crown-Pc), tetrakis[4′,5′-bis(decoxy)benzo18-crown-6]-phthalocyanine, has been prepared [49]. Phthalocyanine (Pc) is a π-conjugated planar molecule, which has been widely used as an electron donor in research on molecular conductors [50]. Partially oxidized Pc molecules form a 1D π–π stack with metallic conducting behavior. Several types of crown-ether-fused Pc molecules have been prepared in order to construct a supramolecular assembly through ion recognition at the crown-ether moieties. The amphiphilic crown-Pc molecule (**1**, Fig. 8.11a) was designed to add amphiphilic character to these crown-ether-fused Pc molecules; it is composed

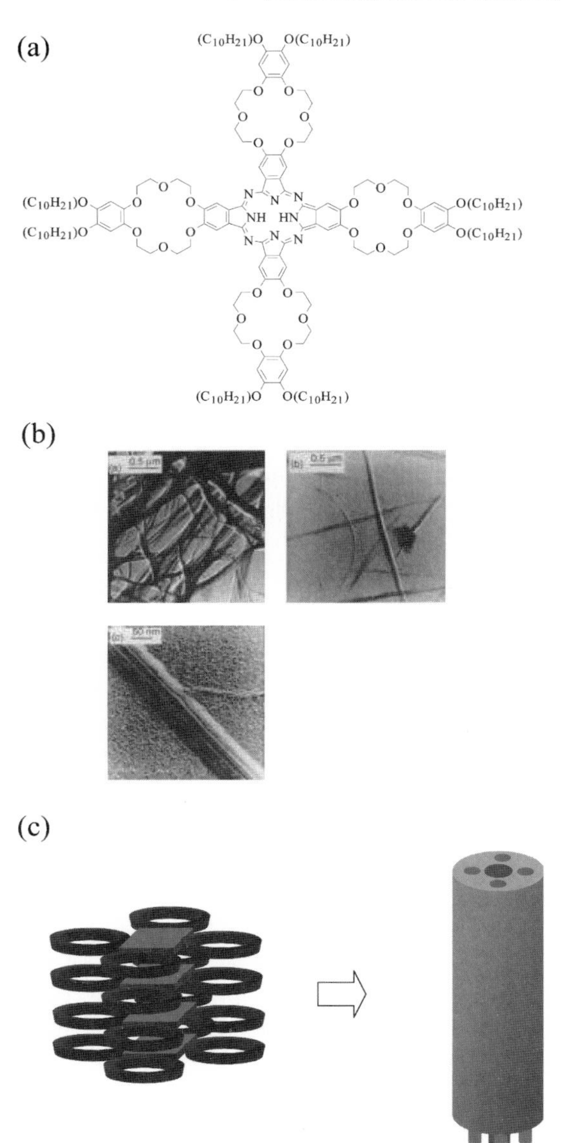

Fig. 8.11. (**a**) Molecular structure of amphiphilic tetrakis[4′,5′-bis(decoxy)benzo18-crown-6]phthalocyanine (**1**). (**b**) TEM images of the gel state of (**1**) in $CHCl_3$ solution. The *scale bars* in the *upper* and *lower parts* correspond to 500 nm and 50 nm, respectively (from C.F. van Nostrum et al., J. Am. Chem. Soc. **117**, 9957 (1995), reprinted with permission). (**c**) Schematic view of assembly structure of (**1**) achieved through the direct stacking of Pc cores and 18-crown-6 rings. The central π–π stack of the Pc core is surrounded by four ionic channels and insulating alkyl chains

of a central Pc core, four 18-crown-6 macrocycles, and eight decyloxy chains linked via benzene rings to the crown ether. The molecule shows a hexagonal columnar mesophase (D_h) in the temperature range from 148 to 178 °C (the decomposition temperature). In the mesophase, the Pc cores are stacked through the π–π interaction and the average interplanar distance of the benzene rings is 0.44 nm.

The molecules (**1**) formed an organogel in chloroform solution. A TEM image of the gel show a dense network of long fibers several micrometers long (Fig. 8.11b). The minimum diameter of the fibers ($\sim$ 6 nm) was consistent with the diameter of the molecule (**1**). The nanowires are composed of up to 10^4 molecules, which are assembled into a 1D wire through π–π stacking. An eclipsed conformation of the π–π interaction of the central Pc core and the outer benzene rings is suggested by electronic spectra of the solution. The central π–π stack of the Pc core may be surrounded by four ionic channels and insulating alkyl chains (Fig. 8.11c). Carrier doping of the molecular nanowire, which is necessary to realize high electrical conductivity, has not been achieved.

8.3.4 TTF Macrocycle

The bis-TTF-substituted macrocycle (**2**, Fig. 8.12) is designed for constructing molecular nanowires, it contains three types of fundamental unit, namely two redox-active TTFs, a macrocyclic polyether, and two decyl chains. These three types of unit are responsible for electrical conduction, ion recognition, and amphiphilic character, respectively [51]. The 1D π–π stacking interaction between the TTF moieties is a driving force for the formation of the 1D nanowire structure, while the macrocyclic polyether part can recognize alkali metal cations in solution. By the application of a Langmuir–Blodgett method to the tetrafluoro-7,7,8,8-tetracyano-*p*-quinodimethane (F_4-TCNQ) complex of **2**, (**2**)$(F_4 - TCNQ)_2$, a CT complex was produced that formed molecular nanowires, which could be oriented on a mica surface by recognizing the cation array at the macrocyclic moiety.

Figure 8.13 shows AFM images (10 $\times$ 10 μm^2 area) of the (**2**)$(F_4\text{-TCNQ})_2$ CT complex transferred from 0.01 M potassium chloride solution. The nano wires had a typical dimension of 2.5 nm (height) $\times$ 50 nm (width) $\times$ 1 μm (length). Figure 8.14 shows selected nanowire structures observed at higher magnification. The characteristic T-shape junctions, parallel nanowire arrays, and nanowire tree structures seen in the figures may be utilized for three-terminal devices, crossbar switches, and network devices, respectively. Mica (muscovite) has a stoichiometry of $KAl_3Si_3O_{10}(OH)_2$, in which the K^+ ions are located in the interlayers of the $Al_3Si_3O_{10}(OH)_2$ hexagonal lattice [52]. The nanowires recognize the cation array on the mica surface during the film deposition process, as a highly oriented nanowire network structure was observed. The electrical conductivity of the nanowires was estimated to be 10^{-3} Scm^{-1}.

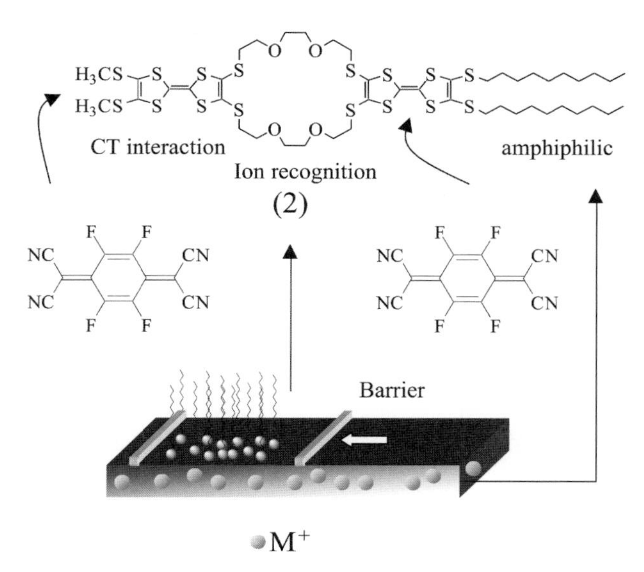

Fig. 8.12. Molecular structure of the amphiphilic TTF macrocycle (**2**). The molecule contains three types of fundamental units, namely two redox-active TTFs, a macrocyclic polyether part, and two decyl chains. Ion recognition is achieved at the air–water interface in a Langmuir–Blodgett trough

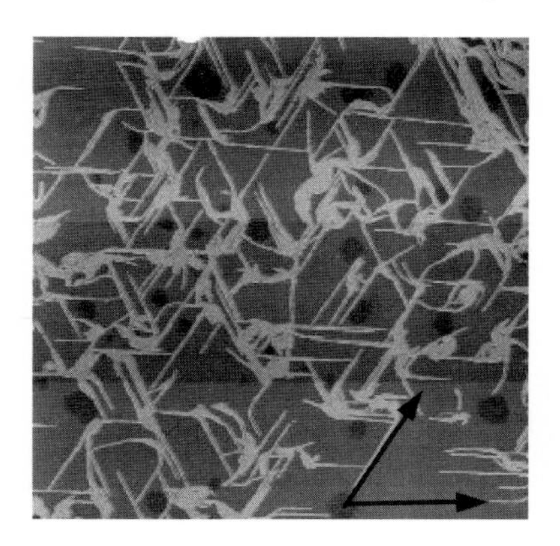

Fig. 8.13. AFM images ($10 \times 10\,\mu m^2$ area) of the (**2**)($F_4 - TCNQ)_2$ CT complex transferred from 0.01 M potassium chloride solution

8.4 Summary

In this chapter, we have described semiconducting and metallic nanowires, both inorganic and organic, which will be important parts of nanoscale electronic devices. The fabrication of devices based on electrically active nanowires will progress according to the following steps: (I) preparation of size (diameter)-controlled nanowires, (II) control of the electronic state (p- or n-type and doping level) of nanowires, (III) orientation control of nanowires on the substrate surface, (IV) fabrication of fundamental device units (diodes,

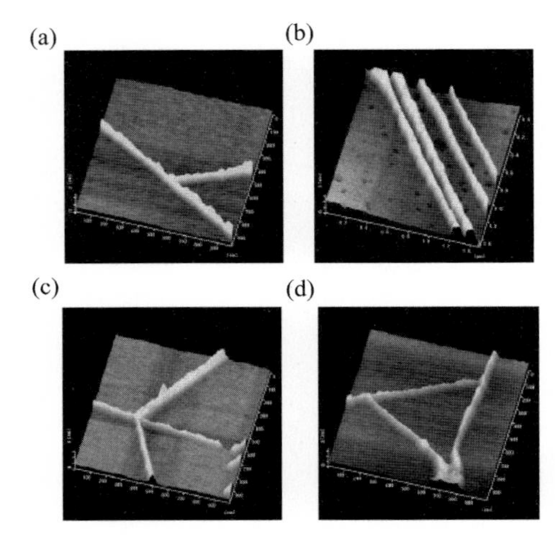

Fig. 8.14. Selected images of molecular nanowire structures. (**a**) T-shape junction, (**b**) parallel nanowire array, (**c**) nanowire tree, and (**d**) nanowire triangle

transistors, etc.) and wiring them to electrodes, and (V) integration of device elements. It is highly desirable to use bottom-up self-assembly processes during the fabrication of nanoscale electronic devices. Top-down nanolithographic techniques will need to be combined with bottom-up processes, and the integration of top-down and bottom-up techniques will open up new possibilities for nanowire-based electronics.

Molecular-assembly nanowires have the potential to allow us to build nanoscale electronic systems by all-self-assembly chemical processes. The design of weak intermolecular interactions (hydrogen-bonding, ion recognition, coordination bonding, hydrophobic–hydrophilic interaction, etc.) to be used in the supramolecular approach will be a key to constructing complex molecular-assembly structures. However, researchers have only just started on the long path to the design and synthesis of programmed molecules which form electroactive molecular nanowires, orient themselves on a substrate surface into a transistor configuration, and self-assemble into integrated molecular circuits. TTF macrocycles are an example of such programmed molecules that form self-assembly nanowires oriented on a substrate. Progress in such chemical design will realize molecular electronics based on molecular nanowires in the near future.

References

1. SEMATECH: *International Technology Roadmap for Semiconductors* (SEMATECH, 1999)
2. P.A. Antón, R. Silberglitt, J. Schneider: *The Global Technology Revolution* (Rand, Santa Monica 2001)
3. F.L. Carter (ed.): *Molecular Electronic Devices* (Dekker, New York 1987); G.J. Ashwell (ed.): *Molecular Electronics* (Wiley, New York 1992); M.C. Petty,

M.R. Bryce, D. Bloor: *Introduction to Molecular Electronic Devices* (Oxford University Press, New York 1995); A. Aviram, M. Ratner (eds.): *Molecular Electronics: Science and Technology* (New York Academy of Science, New York 1998); C. Joachim, J.K. Gimzewski, A. Aviram: Nature **408**, 541 (2000)
4. J.R. Heath, P.J. Kuekes, G.S. Snider, R.S. Williams: Science **280**, 1716 (1998); J.R. Heath: Pure Appl. Chem. **72**, 11 (2000)
5. D.K. Ferry, S.M. Goodnick: in *Transport in Nanostructures*, ed. by H. Ahmed, M. Pepper, A. Broers (Cambridge University Press, Cambridge 1997)
6. B. Albert, D. Bray, A. Johnson, J. Lewis, M. Raff, K. Roberts, P. Walter: *Molecular Biology of the Cell* (Garland, New York 1994); L. Stryer: *Biochemistry* (Freeman, New York 1995)
7. J.G. Nicholls, A.R. Martin, B.G. Wallace, P.A. Fuchs: *From Neuron to Brain*, 4th ed. (Sinauer Associates, New York 2001)
8. J.M. Tour: Acc. Chem. Res. **33**, 791 (2000)
9. A. Tsuda, A. Osuka: Science **293**, 79 (2001); A. Tsuda, A. Osuka: Adv. Mater. **14**, 75 (2002)
10. S.J. Tans, A.R.M. Verschueren, C. Dekker: Nature **393**, 49 (1998); Z. Yao, H.W.C. Postma, L. Balents, C. Dekker: Nature **402**, 273 (1999); M.S. Fuhrer, J. Nygård, L. Shih, M. Forero, Y.G. Yoon, M.S.C. Mazzoni, H.J. Choi, J. Ihm, S.G. Louie, A. Zettl, P.L. McEuen: Science **288**, 494 (2000); A. Bachtold, P. Hadley, T. Nakanishi, C. Dekker: Science **294**, 1317 (2001); M. Ahlskog, R. Tarkiainen, L. Roschier, P. Hakonen: Appl. Phys. Lett. **77**, 4037 (2000); M. Krüger, M.R. Buitelaar, T. Nussbaumer, C. Schönenberger: Appl. Phys. Lett. **78**, 1291 (2001); P.G. Collins, M.S. Arnold, P. Avouris: Science **292**, 706 (2001)
11. J. Hu, T.W. Odom, C.M. Lieber: Acc. Chem. Res. **32**, 435 (1999)
12. X. Duan, C.M. Lieber: Adv. Mater. **12**, 298 (2000)
13. W. Shi, Y. Zheng, N. Wang, C. Lee, S. Lee: Adv. Mater. **13**, 591 (2001)
14. X. Duan, C.M. Lieber: J. Am. Chem. Soc. **122**, 188 (2000); C. Chen, C. Yeh, C. Chen, M. Yu, H. Liu, J. Wu, K. Chen, L. Chen, J. Peng, Y. Chen: J. Am. Chem. Soc. **123**, 2791 (2001); C.C. Tang, S.S. Fan, H.Y. Dang, P. Li, Y.M. Liu: Appl. Phys. Lett. **77,** 1961 (2000)
15. M.S. Gudiksen, C.M. Lieber: J. Am. Chem. Soc. **122**, 8801 (2000); Y. Cui, L.J. Lauhon, M.S. Gudiksen, J. Wang, C.M. Lieber: Appl. Phys. Lett. **78**, 2214 (2001)
16. M.S. Gudiksen, J. Wang, C.M. Lieber: J. Phys. Chem. B **105**, 4062 (2001)
17. Y. Cui, X. Duan, J. Hu, C.M. Lieber: J. Phys. Chem. B. **104**, 5213 (2000)
18. S. Chung, J. Yu, J.R. Heath: Appl. Phys. Lett. **76**, 2068 (2000)
19. Y. Cui, C.M. Lieber: Science **291**, 851 (2001)
20. Y. Huang, X. Duan, Y. Cui, L.J. Lauhon, K. Kim, C.M. Lieber: Science **294**, 1313 (2001)
21. Y. Cui, Q. Wei, H. Park, C.M. Lieber: Science **293**, 1289 (2001)
22. J. Wang, M.S. Gudiksen, X. Duan, Y. Cui, C.M. Lieber: Science **293**, 1455 (2001); X. Duan, Y. Huang, Y. Cui, J. Wang, C.M. Lieber: Nature **409**, 66 (2001); H. Kind, H. Yan, B. Messer, M. Law, P. Yang: Adv. Mater. **14**, 158 (2002)
23. T. Rueckes, K. Kim, E. Joselevich, G.Y. Tseng, C. Cheung, C.M. Lieber: Science **289**, 94 (2000)

24. R.S. Wagner, W.C. Ellis: Appl. Phys. Lett. **4**, 89 (1964); G.A. Bootsma, H.J. Gassen: J. Cryst. Growth **10**, 223 (1971); Y. Wu, P. Yang: J. Am. Chem. Soc. **123**, 3165 (2001)
25. D. Routkevitch, T. Bigioni, M. Moskovits, J.M. Xu: J. Phys. Chem. **100**, 14037 (1996)
26. T.T. Albrecht, J. Schotter, G.A. Kästle, N. Emley, T. Shibauchi, L.K. Elbaum, K. Guarini, C.T. Black, M.T. Tuominen, T.P. Russell: Science **290**, 2216 (2000)
27. B.H. Hong, S.C. Bae, C. Lee, S. Jeong, K.S. Kim: Science **294**, 348 (2001)
28. E. Braun, Y. Eichen, U. Sivan, G. Ben-Yoseph: Nature **391**, 775 (1998); J. Richter, M. Mertig, W. Pompe, I. Mönch, H.K. Schackert: Appl. Phys. Lett. **78**, 536 (2001)
29. T.J. Trentler, K.M. Hickman, S.C. Goel, A.M. Viano, P.C. Gibbons, W.E. Buhro: Science **270**, 1791 (1995); T.J. Trentler, S.C. Goel, K.M. Hickman, A.M. Viano, M.Y. Chiang, A.M. Beatty, P.C. Gibbons, W.E. Buhro: J. Am. Chem. Soc. **119**, 2172 (1997); J.R. Heath, F.K.A. LeGoues: Chem. Phys. Lett. **208**, 263 (1993)
30. J.D. Holmes, K.P. Johnston, R.C. Doty, B.A. Korgel: Science **287**, 1471 (2000)
31. M.P. Zach, K.H. Ng, R.M. Penner: Science **290**, 2120 (2000)
32. A.M. Morales, C.M. Lieber: Science **279**, 208 (1998)
33. J.H. Golden, F.J. DiSalvo, J.M.J. Fréchet, J. Silcox, M. Thomas, J. Elman: Science **273**, 782 (1996)
34. B. Messer, J.H. Song, P. Yang: J. Am. Chem. Soc. **122**, 10232 (2000)
35. Y. Huang, X. Duan, Q. Wei, C.M. Lieber: Science **291**, 630 (2001)
36. P.A. Smith, C.D. Nordquist, T.N. Jackson, T.S. Mayer, B.R. Martin, J. Mbindyo, T.E. Mallouk: Appl. Phys. Lett. **77**, 1399 (2000)
37. M. Batzill, F. Bardou, K.J. Snowdon: Phys. Rev. B **63**, 233408 (2001)
38. S.W. Chung, G. Markovich, J.R. Heath: J. Phys. Chem. B **102**, 6685 (1998)
39. K.D. Hermanson, S.O. Lumsdon, J.P. Williams, E.W. Kaler, O.D. Velev: Science **294**, 1082 (2001)
40. C. Kittel: *Introduction to Solid State Physics*, 6th ed. (Wiley, New York 1986)
41. J.D. Wright: *Molecular Crystals* (Cambridge University Press, Cambridge 1987); E.A. Silinsh, V. Cápek: *Organic Molecular Crystals, Interaction, Localization and Transport Phenomena* (AIP Press, New York 1994)
42. J.M. Lehn: *Supramolecular Chemistry* (VCH, Weinheim 1995); F. Vögtle: *Supramolecular Chemistry* (Wiley, Tokyo 1995); J.W. Steed, J.L. Atwood: *Supramolecular Chemistry* (Wiley, Chichester 2000)
43. T. Akutagawa, T. Hasegawa, T. Nakamura: *Handbook of Advanced Electronic and Photonic Materials and Devices*, Vol 3, ed. by H.S. Nalwa (Academic Press, San Diego 2000)
44. D.O. Cowan: *New Aspects of Organic Chemistry*, ed. by Z. Yoshida, T. Shiba, Y. Oshiro (Kodansha, Tokyo 1989); *Organic Conductors*, ed. by J.P. Farges (Dekker, New York 1994); *Handbook of Organic Conductive Molecules and Polymers*, Vol. 1. ed. by H.S. Nalwa (Wiley, Stuttgart 1997)
45. J.M. Williams, J.R. Ferraro, R.J. Thorn, K.D. Carlson, U. Geiser, H. Wang, A.M. Kini, M.H. Whangbo: *Organic Superconductors* (Prentice-Hall, Englewood Cliffs 1992); T. Ishiguro, K. Yamaji, G. Saito: *Organic Superconductors*, 2nd ed. (Springer, New York 1998)
46. F. Favier, H. Liu, R.M. Penner: Adv. Mater. **13**, 1567 (2001)

47. F.B. Kaufman, E.M. Engler, D.C. Green, J.Q. Chambers: J. Am. Chem. Soc. **98**, 1596 (1976); B.A. Scott, S.J. LaPlace, J.B. Torrance, B.D. Silverman, B. Welber: J. Am. Chem. Soc. **99**, 6631 (1977); J.S. Miller (ed.); *Extended Linear Chain Compounds* (Plenum, New York 1982)
48. J.H. Fuhrhop, J. Köning: in *Membranes and Molecular Assemblies: The Synkinetic Approach*, ed. by J.F. Stoddart (Royal Society of Chemistry, Cambridge 1994)
49. C.F. van Nostrum, S.J. Picken, A.J. Schouten, R.J.M. Nolte: J. Am. Chem. Soc. **117**, 9957 (1995); C.F. van Nostrum, R.J.M. Nolte: Chem. Commun. 2385 (1996); C.F. van Nostrum: Adv. Mater. **8**, 1027 (1996); H. Engelkamp, S. Middelbeek, R.J.M. Nolte: Science **284**, 785 (1999)
50. T.J. Marks: Angew. Chem. Int. Ed. Engl. **29**, 857 (1990)
51. T. Akutagawa, T. Ohta, T. Hasegawa, T. Nakamura, C.A. Christensen, J. Becher: Proc. Natl. Acad. Sci. USA. **99**, 5028 (2002)
52. R.W.G. Wyckoff: *Crystal Structures*, Vol. 4. (Wiley, New York 1960); J. Lima-de-Faria: *Structural Mineralogy: An Introduction* (Kluwer, Dordrecht 1994)

9 Control of Dye Aggregates in Microscopic Polymer Matrices

Olaf Karthaus

Summary. Organic dyes and their aggregates play an important role in many fields of science, ranging from biology to materials science. In the field of molecular electronics, dyes and their aggregates can have various functions, such as energy transfer, electronic conduction, energy transduction, electroluminescence. In order to achieve the desired function, the problems of size control, addressability, and the ability to form complex structures have to be solved.

The regular arrangement of porphyrin molecules in the antenna proteins of photosynthetic bacteria is a beautiful example where the assembly of a few tens of dye molecules in a protein leads to fast and effective energy transfer of sunlight [1,2]. Without the supramolecular, controlled assembly of more than 50 porphyrin molecules contained in at least three different proteins (LH1, LH2, and the photosynthetic reaction center) into a nanometer-sized unit, this special function cannot be realized.

In this chapter, I want to introduce a novel approach to realizing a nanometer-scale arrangement of synthetic dyes and polymers into ordered structures – not unlike the antenna proteins – for possible application in molecular electronics and photonics.

9.1 Cyanine Dyes

Cyanine dyes are an excellent example of dye aggregates in the field of materials science.

In the 1930s Scheibe and Jelly independently reported a deviation from the Lambert–Beer law in solutions of cyanine dyes [3,4]. Above a certain threshold concentration, a narrow redshifted absorption band appeared. Later it was found that this behavior occurs for various types of dyes, and the terms Scheibe-Aggregates and J-aggregates (J for Jelly) were coined to characterize this peculiar state. The narrow absorption band in the visible, which can be fine-tuned by chemical modification of the chromophore, makes J-aggregates ideal as photosensitizers for photographic color films [5,6]. This redshift is due to a strong intermolecular dipole–dipole interaction, and Davidov developed a model in which this dipole–dipole interaction is strongly dependent on the tilt angle of the molecules [7]. Large tilt angles lead to a redshifted absorption, while small tilt angles lead to a blueshifted absorption band [8]. It was also found that J-aggregates of cyanine dyes show a smaller Stokes shift than do the monomer states and hence have narrow, blueshifted fluorescence spectra compared with those of the monomer states.

These peculiar photophysical properties not only led to practical applications, but also attracted theoretical scientists. Even though it seems that the J-aggregate is a peculiar aggregation state with a characteristic absorption and fluorescence behavior, which is different from that of the monomer or of other aggregates, it was predicted that the absorption should depend on the size of the J-aggregate [9–11]. Larger J-aggregates show a slightly stronger redshift in absorption and "higher-quality" aggregates show narrow absorption. Muenter et al. used structural analogues of a mixture of cyanine dyes, to produce J-aggregates with an adjustable statistical mean size of and found that larger J-aggregates have shorter fluorescence lifetimes [12]. Minoshima et al. measured the femtosecond nonlinear optical dynamics of J-aggregates and found that the excitons are confined to one coherent aggregate with a size of ca. 20 molecules [13].

With the development of proximal probes, such as the scanning tunneling microscope and the atomic force microscope in the 1980s it became possible to investigate the structure of J-aggregates on the nanometer scale. In particular, the scanning nearfield optical microscope (SNOM), in which an optical probe with subwavelength dimensions is scanned over a surface [19], enables the resolution of the photonic properties of dye aggregates on a nanometer scale. Higgins et al. used a SNOM to image fibrous J-aggregates embedded in a poly(vinylalcohol) matrix [14] and investigated energy migration in these aggregates [15]. These authors reported that the observation of energy migration was limited by the lateral resolution of the SNOM probe, which was less than 50 nm. This showed that the investigation of single cyanine aggregates was hampered in some way by the fibrous structure of the polymer-embedded dyes.

Hence the need for the preparation of isolated J-aggregates with a defined size at the molecular level.

Research in this field has focused on two approaches to controling the size and orientation of J-aggregates: adsorption from aqueous solution onto muscovite mica [16], and adsorption onto a lipid monolayer at a gas–water interface [17,18]. The results show that well-ordered aggregates with a narrow size distribution can be prepared. But, since the interface is atomically flat and nonstructured, the aggregates are densely packed at the interface, and the optical properties of single aggregates cannot be measured easily.

Here, an alternative method for producing well-separated arrays of J-aggregates on surfaces is proposed: incorporation into microscopic polymer droplets.

9.2 Dewetting-Assisted Polymer Microdome Formation

Dewetting is a spatio-temporal process in which a thin liquid film contracts and forms a random arrangement of polymer droplets with various sizes on a surface. The origin of dewetting stems from the interfacial free energy of the thin film on the surface. If the total energy of the system for the ruptured, dewetted film is smaller than for a spread film, dewetting will take place [20,21]. The total energy of the system must be calculated from all intermolecular interactions, and thus whether dewetting occurs or not depends on the materials (polymer, substrate, and solvent). Dewetting has been intensively studied by Reiter et al. [21,22] and Sheiko et al. [23] and described theoretically by de Gennes [24] for the case of a polymer melt on a flat substrate.

Recently it was found that micrometer-sized polymer droplets, or "domes", can be prepared by casting dilute polymer solutions on substrates [25]. Many different polymers, e.g. polystyrene [26], dendrimers [27], and poly(hexylthiophene) [28], can form various mesoscopic patterns. Typically, the domes in these patterns are 0.5 to several tens micrometers in diameter and five to 500 nm in height. The spatial separation of the domes is of the order of five times their diameter, and thus is at least several micrometers. This fixation on a substrate and spatial separation makes polymer domes ideal matrices to study dye aggregation, and for the spectroscopy of single dye aggregates.

Since diffusion of molecules is restricted to one single dome (diffusion of molecules between domes cannot take place, since they are spatially separated), we are able to control the number of molecules of any given additive in each dome.

For J-aggregates this means that by adjusting two parameters, the dome size and the dye concentration, we should be able to control the aggregate size. With increasing dome size and concentration, the number of dye molecules per dome increases. Hence the potential size of the J-aggregate increases accordingly.

9.3 Results

9.3.1 Microscopy of Polystyrene Matrices

A cast droplet of 100 μl volume will take a few minutes to evaporate under ambient conditions. In situ optical microscopy (Fig. 9.1) shows that a fingering instability at the edge of the solution droplet is responsible for the deposition of micrometer-sized polymer domes on the substrate. Under appropriate conditions, regular arrays of equally-sized domes can be obtained, although this is not required for the purpose of this research.

By adding aliquots of **1** (Fig. 9.2) to a polystyrene solution, samples with concentrations from 1 ppm to 10 wt % dye/polystyrene were prepared.

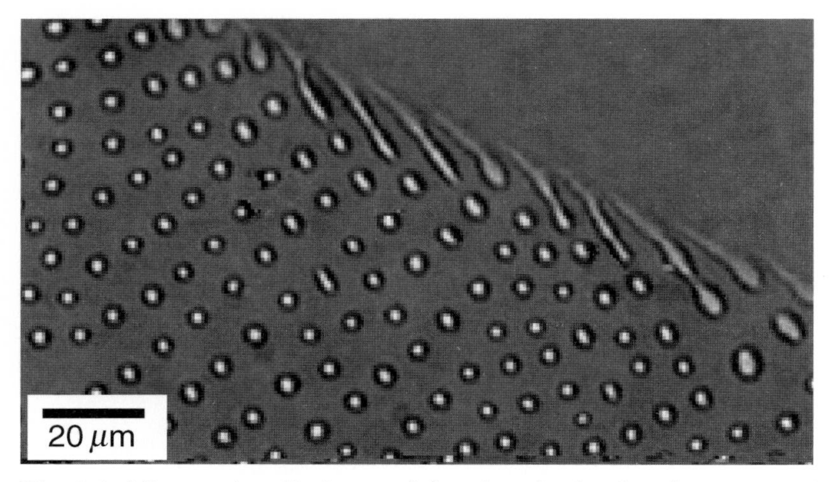

Fig. 9.1. Microscopic still picture of the edge of a droplet of an evaporating chloroform solution containing 1 mg/ml polystyrene

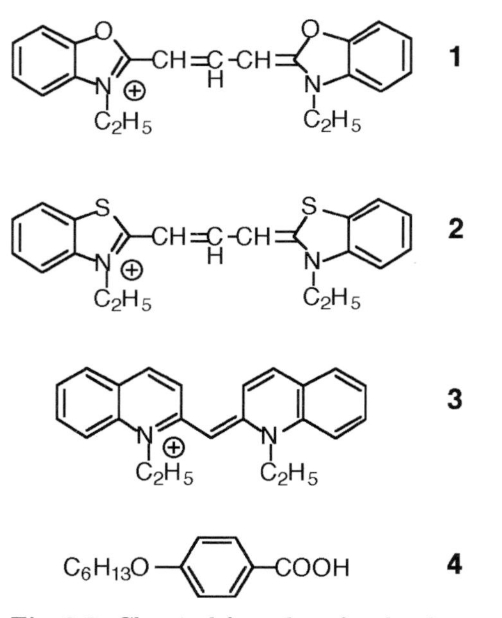

Fig. 9.2. Chemical formulae of molecules used here

Figure 9.3 shows micrographs for a series of concentrations. The optical micrographs of higher-concentration samples (a, 10%, and b, 1%) show a texture within the polymer domes, which implies that macroscopic phase separation within each polymer dome has taken place. The darker, round features within the domes are presumably larger dye aggregates, or crystallites. Fluorescence

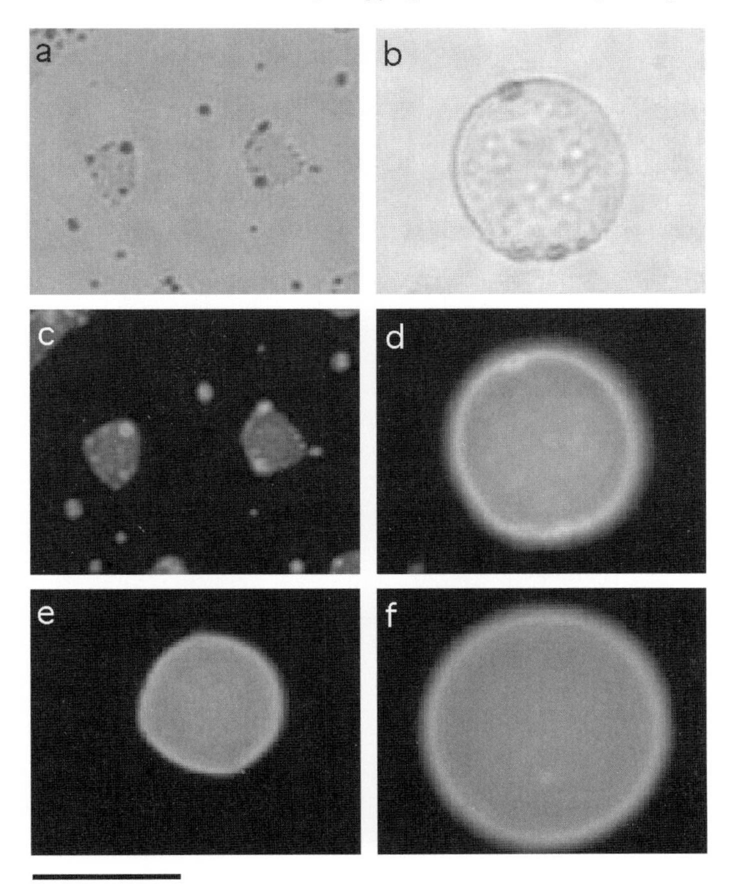

Fig. 9.3. Optical (**a** and **b**) and fluorescence micrographs (**c**–**f**) of polystyrene domes with different concentrations of cyanine dye **1**: (**a**) and (**c**), 10%; (**b**) and (**d**), 1%; (**e**), 0.1%; (**f**), 0.01%. The *bar* corresponds to 10 μm

images taken at the same positions (Figs. 9.3c, d) reveal the presence of two different fluorescing species, which emit green and orange fluorescence, respectively. The presence of such large aggregates is not surprising, since the ionic dye is embedded in a hydrophobic matrix of polystyrene, a hydrocarbon.

Lower concentrated samples with 0.1 and 0.01% of dye show only green fluorescence (Figs. 9.3e, f). The corresponding optical micrographs have been omitted, since no phase separation was detected. Instead, the polystyrene domes were completely transparent. The fluorescence was homogeneously distributed within the polymer domes. Since the domes have a larger thickness in the center than at the edge, this finding is unexpected, but can explained if the possibility of adsorption of the cationic dye onto the anionic glass substrate is taken into account. If this occurred, the dye concentration would be nearly independent of the thickness of the polymer dome.

Other cyanine dyes, not only those with a similar structure, such as **2** (Fig. 9.2), but also those with a different chemical structure, such as **3**, show a similar behavior. Higher-concentration dyes phase-separate and form large aggregates, which give contrast in optical microscopy. Lower-concentration samples give homogeneous fluorescence.

9.3.2 Spectroscopy

In order to obtain a deeper insight into the spectroscopic properties of the dye aggregates and the polymer domes, fluorescence spectromicroscopy was undertaken. The experimental setup allowed the collection of the fluorescence light from a spot 3 μm in diameter. Since the spacing between polymer domes was larger than 3 μm, the spectra of single domes could be measured. Figure 9.4 shows the fluorescence spectra of single polymermicrodomes which contain 1 wt % of **1**, **2**, and **3**.

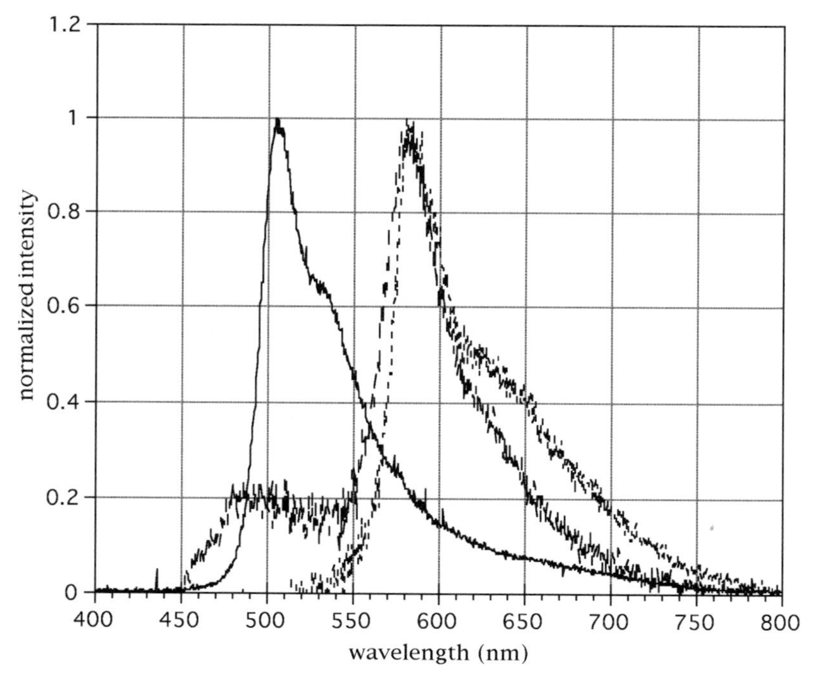

Fig. 9.4. Normalized fluorescence spectra of **1** *(solid line)*, **2** *(dotted line)*, and **3** *(broken line)*. Excitation wavelength: 400–440 nm. For each spectrum, the background fluorescence was subtracted

Backscattered excitation light (400 – 440 nm) was cut off below 460 nm, which is well separated from the peak maxima of all the dyes. Thus, the shape of the peaks was not influenced by the filter. All spectra show a narrow fluorescence peak with a width of less than 20 nm at 80% intensity. This, and the peak positions, are characteristic of J-aggregates.

In the following, the spectral characteristics of two of the dyes will be examined in more detail.

N, N′-diethyloxacarbocyanine, **1**. Figure 9.5 shows the fluorescence spectra of **1** at three different concentrations in polystyrene domes, and in addition a reference spectrum obtained from a poly(vinyl alcohol) cast film containing 10 wt % dye. The reference sample was prepared according to the literature, by mixing hot solutions of polymer and dye and immediate casting. The emission maximum is at 504 nm and has a width of 14 nm at 80% intensity. A redshifted shoulder in the fluorescence spectrum may stem from residual monomer fluorescence, but its origin is not unambiguously clear.

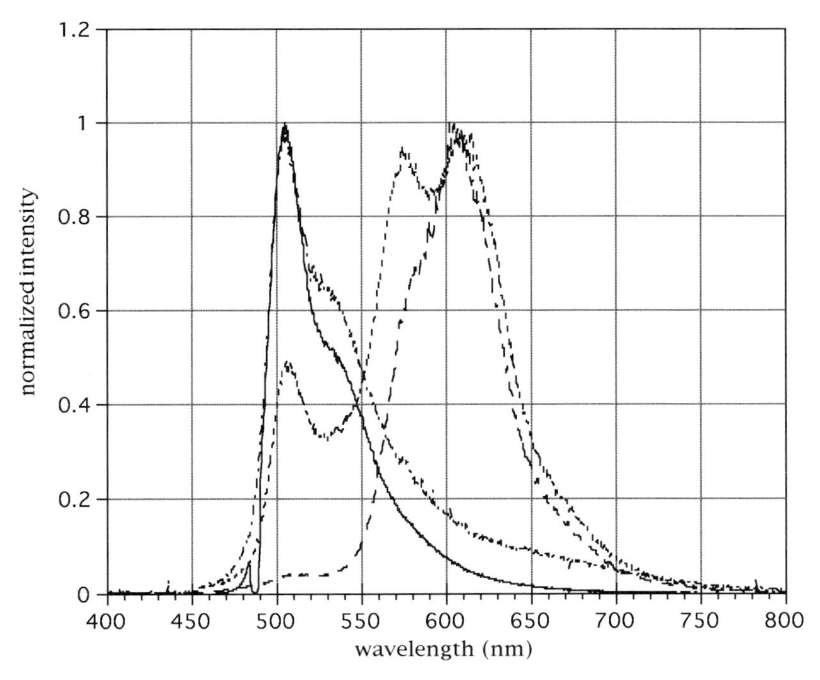

Fig. 9.5. Fluorescence spectra of **1** in various environments: 10 wt % in poly(vinyl alcohol) *(solid line)*, 0.1% in polystyrene *(broken–dotted line)*, 1% in polystyrene *(dotted line)*, and 10% in polystyrene *(broken line)*. The poly(vinylalcohol) sample was excited with an Ar ion laser (488 nm). The polystyrene samples were excited with a 200 W Hg lamp (400 – 440 nm).

In comparison with the reference spectrum, the dilute sample of **1** in polystyrene has exactly the same characteristics, except that the shoulder at 550 nm has a slightly higher intensity. This is a strong indication of the presence of J-aggregates.

Higher-concentration samples of **1** in polystyrene show a totally different behavior. 10 wt % of dye yields polymer domes with large aggregates (see Figs. 9.3a, c). Fluorescence spectra show a broad fluorescence peak at around 610 nm. The width at 80% intensity is at least 40 nm, and together with the microscopic data it can be concluded that these samples contain 3-D crystals. Less concentrated samples with 1% dye show nearly homogeneous fluorescence (Figs. 9.3b, d), but the fluorescence spectra still show a predominantly crystalline sample, even though the onset of the aggregate fluorescence at 505 nm can be seen.

Spectroscopy over a large, macroscopic sample will average out the fluorescence spectra of many dye aggregates. On the other hand, microspectroscopy offers the advantage of measuring the local spectral properties of a single dye aggregate, or at most a few. Hence information about the local environment, which may vary even in an apparently homogeneous sample, can be obtained.

Figure 9.6 shows the fluorescence spectra of three different polymer domes

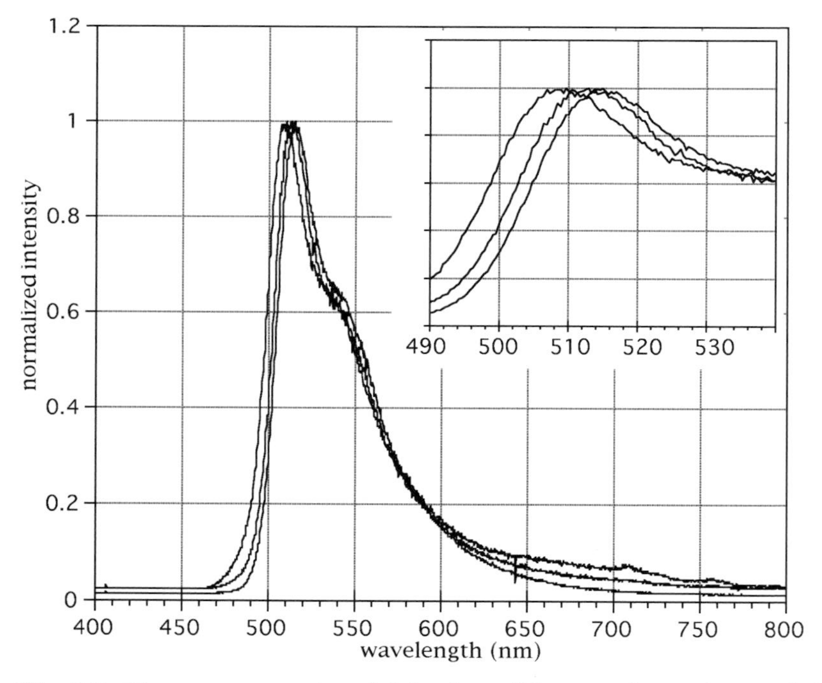

Fig. 9.6. Fluorescence spectra of **1** in three different polymer domes of the same sample (1%). The *(inset)* is a detailed view of the peak maxima

in the same sample (1% **1**). Three domes with approximately the same fluorescence intensity were recorded. Even though the shapes of the spectra are very similar, a clear difference in the wavelength of the fluorescence maximum, which ranges from 508 to 514 nm, can be seen. Even though the difference is only a few nanometers, it is significant and is most likely due to a difference in aggregate size.

3,3′-diethyl-isocyanine, **3**. This isocyanine compound has been already been thoroughly investigated in detail by other research groups [14–16]. The standard procedure for the formation of J-aggregates is incorporation into a poly(vinyl alcohol) matrix at high dye concentration, typically 10 wt %. In order to verify the performance of our experimental setup for the measurement of fluorescence spectra, such a sample was prepared, in addition to samples in polystyrene matrices. The poly(vinyl alcohol) sample was prepared as described in the literature [14], and the resulting cast film was transparent in the optical microscope. Fluorescence microscope spectroscopy shows a typical J-aggregate with a fluorescence maximum at 582 nm and a width of a mere 6 nm at 80% intensity (Fig. 9.7). The shoulder at around 550 nm can be attributed to residual monomer fluorescence. The absorption

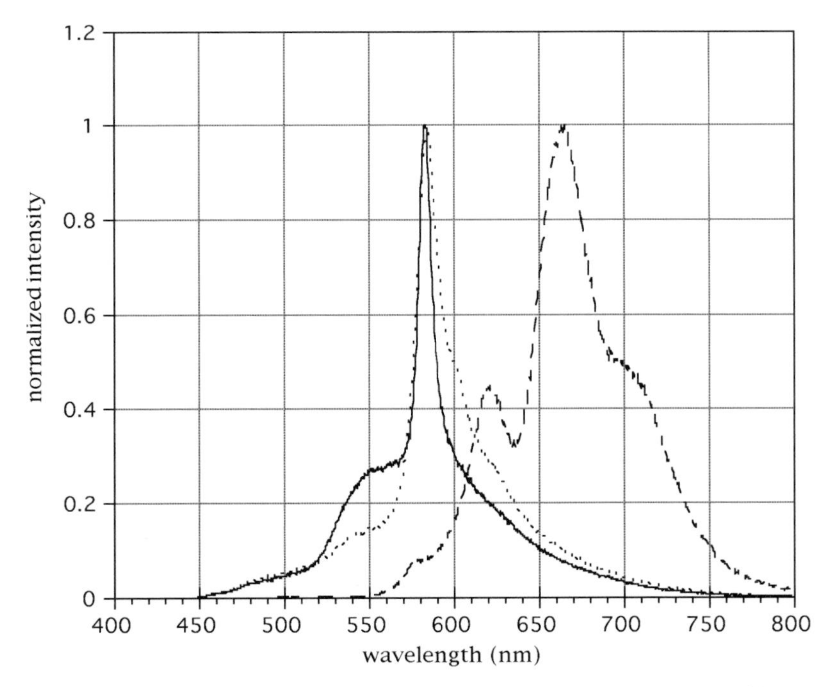

Fig. 9.7. Fluorescence spectra of **3** in various environments: 10 wt % in poly(vinyl alcohol) *(solid line)*, 1 wt % in Polystyrene *(broken line)*, and 50 wt % in polystyrene *(dotted line)*. The samples were excited with a 200 W Hg lamp (400 – 440 nm)

spectrum (data not shown here) has a maximum at 581 nm. This small Stokes shift is a strong indicator of J-aggregation. A sample containing 1% of **3** in polystyrene shows the same characteristic fluorescence (emission maximum at 584 nm, width at 80% intensity of 8 nm), except that the shoulder is now redshifted and appears at around 600 nm. As in the case of **1**, the higher-concentration, crystalline samples have totally different fluorescence spectra, with broad peaks between 600 and 700 nm.

9.3.3 Microscopy of Liquid Crystalline Matrices

Even though it was recently reported that cyanine dyes form liquid-crystalline J-aggregates in an environment [30,31], the influence of a Liquid-crystalline environment on J-aggregate formation has not been reported previously. Compared with amorphous matrices, such as poly(vinyl alcohol) or polystyrene, a nanoscopically ordered environment, such as that of a crystal or liquid crystal, should have an effect on the aggregate formation of an added dye. This additional type of order within each dewetted dome will give rise to a hierarchical environment (the nanometer order of liquid-crystalline molecules and the micrometer order of the dewetted domes), and novel properties can be expected.

On the basis of previous work with polymeric liquid crystals [29], it was found that low-molar-mass liquid-crystalline compounds also dewet and form microscopic domes on substrates. In this study, a simple compound, p-hexyloxy benzoic acid, was used, which has a nematic phase between 74 °C and 113 °C. Casting from a solution which contained 100 ppm to 1% of **1**, and subsequent annealing of the samples at 80 °C for 2 hours led to dewetting and dome formation.

The as-cast sample does not dewet completely, and consists mainly of a continuous film. Annealing leads to a rupture of the film and the formation of micron-sized domes. Figure 9.8 shows fluorescence images of a sample an-

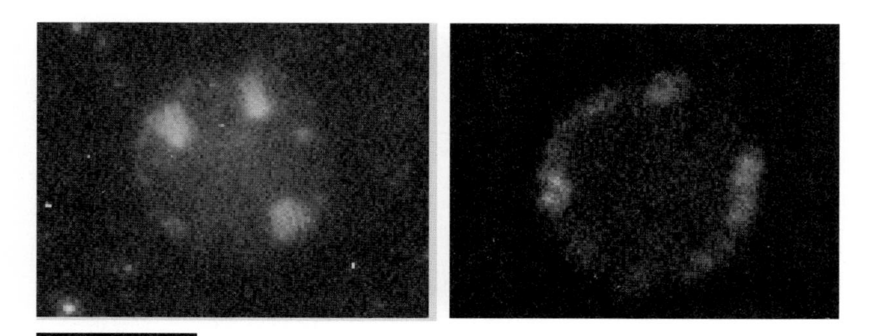

Fig. 9.8. Fluorescence micrographs of cyanine dye **1** in a matrix of **4** (0.1 wt %), annealed at 80 °C (**a**) and 140 °C (**b**). The *bar* corresponds to 10 μm

nealed at 80 °C and 140 °C. Even though the morphology – small fluorescent areas within the matrix dome – is similar for both samples, the fluorescence colors are different. The fluorescence of the high-temperature sample is bright green, whereas the low-temperature annealed sample shows a less bright, green–yellowish fluorescence.

Figure 9.9 shows the corresponding fluorescence spectra of an as-cast film, after annealing at 80 °C (that is, in the nematic phase) and after annealing at 140 °C (that is, in the isotropic state). Since in situ measurement of the fluorescence spectra is not possible with the present setup, the samples have had to be measured at room temperature and thus in the crystalline state. Even though the crystallization may have an additional effect on the dye aggregation, it became clear that the annealing temperature (in the temperature range of the nematic phase or of the isotropic phase) plays an important role in the dye aggregate formation.

The fluorescence spectrum of the as-cast sample is dominated by two broad peaks, one around 505 nm and one around 570 nm. Annealing in the liquid- crystalline phase leads to an increase of the 505 nm emission in comparison with the 570 nm emission. Annealing at 140 °C gives the most drastic

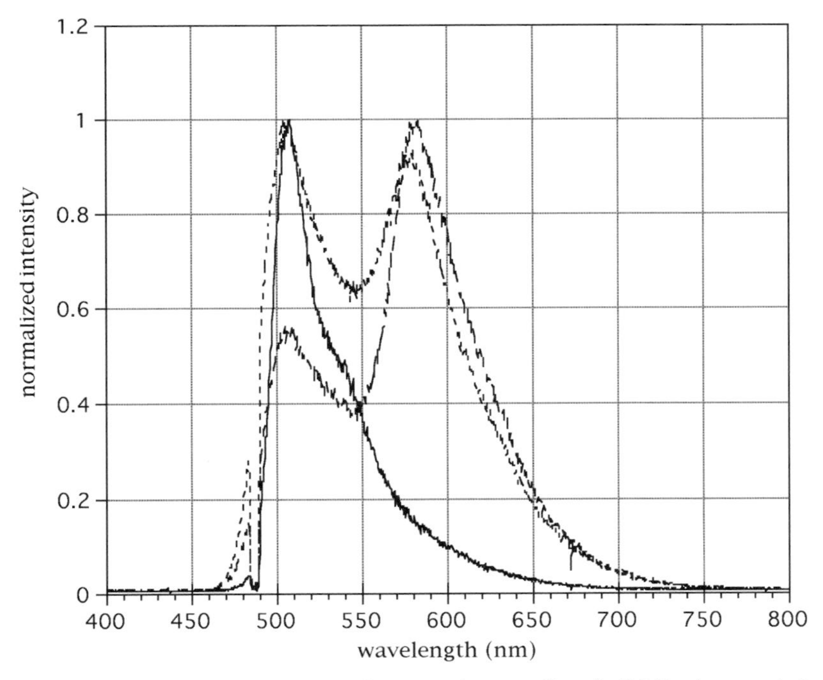

Fig. 9.9. Fluorescence spectra of **1** in **4**: As-cast film *(solid line)*, annealed at 80 °C *(dotted line)*, annealed at 140 °C *(broken line)*. The samples were excited with an Ar ion laser (488 nm)

change in fluorescence emission. In this case, a narrow single peak with spectroscopic features similar to those for the poly(vinyl alcohol) and polystyrene matrices has been obtained. For each sample, the spectra of 14 individual domes is measured. In the case of the as-cast sample and the sample annealed at 80 °C a certain variation of the relative peak heights at 505 and 570 nm has been observed.

These results show that aggregate formation can be controlled not only by the size of a matrix dome or by the dye concentration, but also by other parameters. Since annealing is a process which can be performed at any time after the sample has been prepared, such a change in aggregation may be useful for applications such as data storage or data processing.

9.4 Outlook

Molecular electronics depends on tailor-made functions of various organic materials. Besides the purely electronic functions of electronic conduction, switching, and storage, the interfacing of molecular-electronics components with other devices, e.g. for data input and output, has to be realized. Fluorescent dyes and their aggregates not only play an important role in energy and electron migration processes, but also can be used for data storage and display purposes. For these purposes robust, but also small – if possible at the molecular level – functional dye aggregates have to be constructed.

The approach of incorporating cyanine dyes and their aggregates into micrometer-sized polymer droplets, or "domes" on surfaces is promising for the use of these dyes in molecular electronics.

References

1. G. McDermott, S.M. Prince, A.A. Freer, A.M. Hawthornthwaite-Lawless, M.Z. Papiz, R.J. Cogdell, N.W. Issacs: Nature **374**, 517 (1995)
2. J. Koepke, X. Hu, C. Muenke, K. Schulten, H. Michel: Structure **4**, 58 (1996)
3. E.E. Jelly: Nature **138**, 1009 (1936)
4. G. Scheibe: Angew. Chem. **49**, 563 (1936)
5. T. Tani: in *J-Aggregates*, ed. by T. Kobayashi (World Scientific, Singapore 1996), p. 209
6. W. West, P.B. Gilman: *The Theory of the Photographic Process*, ed. by T.H. James (Macmillan, New York 1977), p. 63
7. A.S. Davidov: *Theory of Molecular Excitons*, (Plenum, 1971)
8. M. Kasha: Radiat. Res. **20**, 55 (1963)
9. D. Möbius, H. Kuhn: Isr. J. Chem. **18**, 375 (1979)
10. D. Möbius, H. Kuhn: J. Appl. Phys. **91**, 683 (1989)
11. F.C. Spano, J.R. Kulinski, S.J. Mukamel: J. Chem. Phys. **94**, 7534 (1991)
12. A.A. Muenter, D.V. Brumbaugh, J. Apolito, L.A. Horn, F.C. Spano, S. Mukamel: J. Phys. Chem. **96**, 2783 (1992)

13. K. Minoshima, M. Taiji, K. Misawa, T. Kobayashi: Chem. Phys. Lett. **218**, 67 (1994)
14. D.A. Higgins, P.J. Reid, P.F. Barbara: J. Phys. Chem. **100**, 1174 (1996)
15. D.A. Higgins, J. Kerimo, D.A. Vanden Bout, P.F. Barbara: J. Am. Chem. Soc. **118**, 4049 (1996)
16. S. Sugiyama, H. Yao, O. Matsuoka, R. Kawabata, N. Kitamura, S. Yamamoto: Chem. Lett. 37 (1999)
17. S. Kirstein, H. Möhwald: J. Chem. Phys. **103**, 826 (1995)
18. S. Kirstein, H. Möhwald, M. Shimomura: Chem. Phys. Lett. **154**, 303 (1989)
19. D.W. Pohl, W. Denk, M. Lanz: Appl. Phys Lett. **44**, 651 (1984)
20. A. Faldini, R.J. Composto, K.I. Winey: Langmuir **11**, 4855 (1995)
21. G. Reiter, P. Auroy, L. Auvray: Macromolecules **29**, 2150 (1996)
22. G. Reiter: Langmuir **9**, 1344 (1993)
23. S. Sheiko, E. Larmann, M. Möller: Langmuir **12**, 4015 (1996)
24. P.-G. de Gennes: Rev. Mod. Phys. **57**, 827 (1985)
25. O. Karthaus, L. Grasjo, N. Maruyama, M. Shimomura: Chaos **9**, 308 (1999)
26. O. Karthaus, K. Ijiro, M. Shimomura: Chem. Lett. 821 (1996)
27. J. Hellmann, M. Hamano, O. Karthaus, K. Ijiro, M. Shimomura, M. Irie: Jpn. J. Appl. Phys. **37**, L816 (1998)
28. O. Karthaus, T. Koito, M. Shimomura: Mater. Sci. Eng. C **8–9**, 523 (1999)
29. O. Karthaus, H. Yabu, K. Akagi, M. Shimomura: Mol. Cryst. Liq. Cryst. **364**, 395 (2001)
30. W.J. Harrison, D.L. Mateer, G.J.T. Tiddy: Faraday Discuss. **104**, 139 (1996)
31. H. Stegemeyer, F. Stockel: Ber. Bunsenges. **100**, 9 (1996)

Part III

Theory of Nanomolecular Systems

10 Theoretical Calculations of Electrical Properties of Nanoscale Systems Under the Influence of Electric Fields and Currents

Satoshi Watanabe

Summary. Recently, extensive experimental attempts to clarify the properties of various nanomolecular systems have been performed. Although many interesting results have already been reported and more are expected in the near future, interpretation of the results of such experiments is not necessarily straightforward, because of the strong interaction between the nanomolecular systems and the experimental probe. In this situation, reliable theoretical calculations are useful for interpreting experimental results, exploring novel properties of nanomolecular systems, and deriving guiding principles for designing nanomolecular systems so that they will be optimal for given purposes.

On the other hand, nanomolecular systems present a challenge to theoretical calculations, because conventional methods, which are powerful for examining the properties of bulk crystals and isolated molecules, are often insufficient for nanomolecular systems. First, it is often desirable to consider semi-infinite electrodes explicitly instead of representing them by finite clusters or thin slabs. Second, external fields often play an essential role in measurements of properties of nanomolecular systems, but the conventional methods have serious difficulties in treating their effects.

In this chapter, keeping the above in mind, we review recent theoretical approaches to examining the properties of nanomolecular systems, focusing on those which take account of semi-infinite electrodes and external electric fields explicitly. In particular, we discuss the electrical properties of systems under the influence of electric fields and currents, because these properties are very important in designing molecular electronic devices. After summarizing the problems of the conventional methods in Sect. 10.1, methods developed to examine the electrical properties of nanoscale systems are reviewed in Sect. 10.2. Recent applications of such methods to simple nanoscale systems without organic molecules are then described in Sect. 10.3. To examine such systems is useful, because some of the essential features of the properties of nanomolecular systems are also seen in them. Next, recent progress in theoretical analysis of the electrical properties of nanomolecular systems is reviewed in Sect. 10.4, followed by some concluding remarks.

10.1 Problems of Conventional Methods

Among the methods to examine the electronic and atomic properties of materials, molecular-orbital theory and density functional theory have been utilized most widely. These theories can describe various properties of materials well, in spite of the fact that the effects of electron correlation are not fully taken into account in these theories.

One of the difficulties in the application of these conventional methods to nanomolecular systems is caused by the fact that it is very important to consider not isolated nanomolecules but nanomolecules combined with electrodes and/or probes to predict the results of experimental measurements. When electrodes are attached, the properties of nanomolecular systems are affected significantly. Though such cases can be treated in the conventional methods by representing a semi-infinite electrode by a finite cluster or thin slab, and this treatment is known to be able to describe the atomic and electronic structures of solid surfaces reasonably well in many cases, the validity of this approach is not guaranteed in the case of nanomolecular systems.

In particular, these models involve a serious problem when the effects of electric fields and currents are considered. When an electric field is applied to metal electrodes, charges are induced on their surfaces. In the case of semi-infinite electrodes, such induced charges are supplied from deep inside the electrodes. On the other hand, finite clusters and thin slabs do not have such a charge reservoir, and thus the appearance of an induced charge involves artificial charge redistribution inside them. Because of this, models including semi-infinite electrodes are strongly desired in situations of strong electric fields and currents.

The problems of the conventional methods are not limited to the above. In measurements of electrical properties of nanomolecular systems, the local strength of the electric fields applied to the nanomolecular system may reach a value of the order of 1 to 10 eV/nm, which is much stronger than that for macroscopic systems. Since such strong fields deform the electronic states of the system significantly, the validity of conventional molecular-orbital theory, where the electronic states are described using linear combinations of atomic orbitals, is highly questionable. Further, this treatment cannot describe the field emission of electrons, which often occurs under strong electric fields.

One may think that this problem could be avoided by expanding electronic states using plane waves instead of atomic orbitals. However, this approach does not work either in the cases with strong fields. Since the potential due to a uniform electric field does not satisfy the periodic boundary conditions that are necessary in a plane wave expansion, the potential due to an applied uniform electric field is usually replaced by a sawtooth potential in this approach. When the electric field is strong, this sawtooth potential involves artificial bound states, as illustrated in Fig. 10.1. If the field is strong enough to make the energy of these artificial states lower than the chemical potential of the system, unrealistic charge transfer to the bound states occurs, because only one chemical potential is defined for the whole system in the conventional treatment. This unrealistic charge transfer involves an inappropriate description of the electronic states of the system. It is worth mentioning that similar unrealistic charge transfer occurs regardless of the type of basis

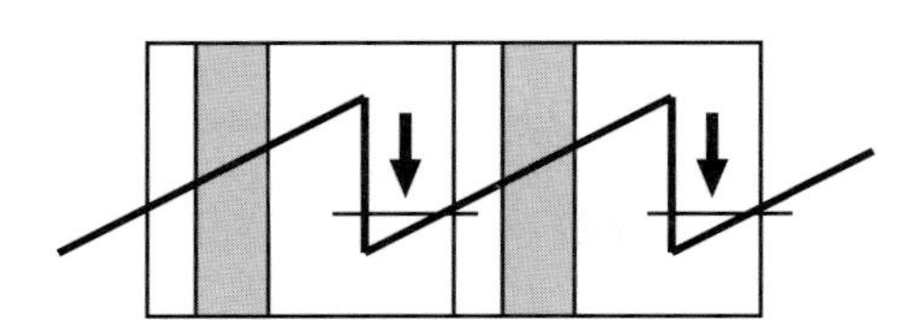

Fig. 10.1. Schematic illustration of artificial bound states caused by a saw-tooth potential. Two supercells are shown. The *shaded regions* denote the regions where the nanomolecular systems are located. Artificial bound states are indicated by *arrows*

functions, when a system consists of more than one object and the electric currents between the objects are negligible.

In cases where electric currents flow, another problem arises. In such cases, any deficiency or excess of charge in an electrode which is caused by the current flow should be compensated immediately from the charge reservoir inside the electrode, resulting in preservation of the difference between the chemical potentials of the electrodes. However, the conventional methods force the chemical potentials of the electrodes to be equal, and thus result in an unrealistic charge distribution again.

For these reasons, the conventional methods have many serious difficulties when they are applied to nanomolecular systems, in particular in the presence of strong electric fields and currents. Therefore, it is very important to develop new methods that are applicable to investigations of the electrical properties of nanomolecular systems under the influence of strong electric fields and currents.

10.2 Methods to Treat Nanomolecular Systems Under the Influence of Electric Fields and Currents

In this section, we review the methods developed to examine the electronic and electrical properties of nanomolecular systems and other nanoscale systems under the influence of strong electric fields and currents, including those developed in our group. First, a method suitable for cases where the system consists of more than one object and electric currents between the objects are negligible is described in Sect. 10.2.1. Next, methods which can also treat cases with current flows are described in Sect. 10.2.2.

10.2.1 Method for Cases Without Electric Currents

To treat nanoscale systems consisting of more than one object under the influence of a strong electric field but with a negligible current, a new method named the partitioned real-space density functional (PRDF) method has been developed recently [1]. Here, we explain this method, taking the case of two objects as an example.

In the PRDF method, we consider a box containing the two objects. This box is sufficiently large that the electronic wave functions can be taken as zero at all the borders of the box. We divide the box into two regions, which we call regions I and II hereafter, between the two objects. Then, we assume that the wave functions are also zero at the boundary between these regions. Since we consider the case where the electric current between the objects can be neglected, this assumption is reasonable. In the PRDF method, the Kohn–Sham equation for the electronic wave functions in region I is solved with the constraint that the wave functions are zero at all the boundaries, including the one between the two regions. In doing so, the electrostatic potential due to charges in region II is taken into account. The wave functions in region II are obtained similarly. In this way, we can avoid the unrealistic electron transfer between the two objects that inevitably occurs in the conventional methods. In practical calculations, the conventional real-space density functional method [2,3] has been adopted, because the constraint that the wave functions are zero at all the boundaries can be incorporated easily into this method. However, it is possible to combine the basic idea of the PRDF method with other methods, such as molecular-orbital theory.

Figure 10.2 shows the equipotential surfaces around two pyramidal jelliums in a uniform external electric field of 1 V/nm calculated by the conventional density functional (DF) method and the PRDF method [1]. In this calculation, the two pyramids are separated from each other by 2 nm, and a rectangular box of dimensions 5.71 nm × 4.63 nm × 4.63 nm is employed. In the case of the conventional DF method (Fig. 10.2a), the electric field is weakened significantly in the region between the two pyramids, while it is enhanced on their outer sides. These features are inconsistent with what simple electrostatics tells us. On the other hand, in the case of the PRDF method (Fig. 10.2b) the features of the field profile are quite reasonable in the light of electrostatics: the field is enhanced at the apexes of the two pyramids. These results demonstrate the difficulties of the conventional methods in

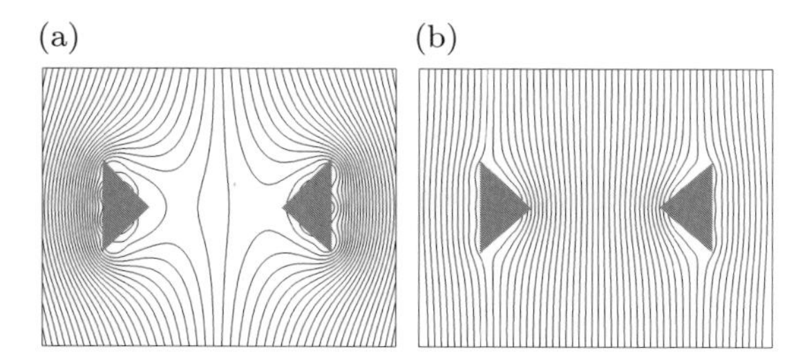

Fig. 10.2. Equipotential surfaces around two pyramidal jelliums calculated by (**a**) the conventional and (**b**) the partitioned real-space density functional method. The pyramids are separated from each other by 2 nm, and are in an electric field of 1 V/nm in the lateral direction. The contour spacing is 0.1 V. From ref. [1]

treating electronic states in the presence of electric fields, together with the effectiveness of the PRDF method.

10.2.2 Methods for Cases with Electric Currents

When one wants to examine the electronic and electrical properties of nano-molecular systems with electric currents, and is not satisfied with a qualitative description using perturbative approaches such as those widely adopted to calculate scanning tunneling microscopy images [4,5], it is necessary to develop methods beyond the one described in Sect. 10.2.1. Several methods have already been proposed for this purpose. They are categorized into two groups, methods using Green's functions and methods based on connection of wave functions. Here, we explain both types of methods briefly, with several miscellaneous remarks.

Methods Using Green's Functions. Briefly speaking, in this approach, the electronic wave functions of a whole system consisting of nanoscale objects and electrodes are obtained, utilizing the Green's functions and wave functions for bare electrodes. The equations to determine the wave functions of the whole system can be expressed in the Lippmann–Schwinger form as follows [6]:

$$\Psi^{\mathrm{EO}}(\boldsymbol{r}) = \Psi^{\mathrm{E}}(\boldsymbol{r}) + \int \mathrm{d}^3\boldsymbol{r}'\, \mathrm{d}^3\boldsymbol{r}''\, G^{\mathrm{E}}(\boldsymbol{r},\boldsymbol{r}')\, \delta V(\boldsymbol{r}',\boldsymbol{r}'')\, \Psi^{\mathrm{EO}}(\boldsymbol{r}'') \,. \quad (10.1)$$

Here, the superscripts E and EO refer to the bare electrodes and the whole system, respectively, $\delta V(\boldsymbol{r},\boldsymbol{r}')$ is the difference between the potentials of the whole system and of the bare electrodes, and $G^{\mathrm{E}}(\boldsymbol{r},\boldsymbol{r}')$ denotes the Green's function of the bare electrodes. Once $G^{\mathrm{E}}(\boldsymbol{r},\boldsymbol{r}')$ is known, $\Psi^{\mathrm{EO}}(\boldsymbol{r})$ can be determined using (10.1), by a recursive procedure starting from an appropriate initial guess for $\Psi^{\mathrm{EO}}(\boldsymbol{r})$ on the right-hand side of (10.1).

The wave functions contributing to the electric current between the electrodes, $\Psi^{\mathrm{EO}}(\boldsymbol{r})$, can be classified into two groups. Let us define the direction perpendicular to the surfaces of the electrodes as the x axis, the electrode that includes $x = -\infty$ as the left electrode, and the electrode that includes $x = +\infty$ as the right electrode. Then one group of wave functions has the asymptotic form

$$\Psi^{\mathrm{EO}}_{\rightarrow}(\boldsymbol{r}) \propto \begin{cases} \mathrm{e}^{\mathrm{i}k_{\mathrm{L}}x} + r\,\mathrm{e}^{-\mathrm{i}k_{\mathrm{L}}x}, & x \to -\infty \\ t\,\mathrm{e}^{\mathrm{i}k_{\mathrm{R}}x}, & x \to +\infty \end{cases} , \quad (10.2)$$

while the other group has the form

$$\Psi^{\mathrm{EO}}_{\leftarrow}(\boldsymbol{r}) \propto \begin{cases} \mathrm{e}^{-\mathrm{i}k_{\mathrm{R}}x} + r\,\mathrm{e}^{\mathrm{i}k_{\mathrm{R}}x}, & x \to +\infty \\ t\,\mathrm{e}^{-\mathrm{i}k_{\mathrm{L}}x}, & x \to -\infty \end{cases} . \quad (10.3)$$

Here, $k_{\rm L}$ and $k_{\rm R}$ satisfy $(1/2)\,k_{\rm L}^2 = E - (1/2)\,|K_{\|}|^2 - v_{\rm eff}^{\rm E}(x = -\infty)$ and $(1/2)\,k_{\rm R}^2 = E - (1/2)\,|K_{\|}|^2 - v_{\rm eff}^{\rm E}(x = +\infty)$, respectively, where E, $K_{\|}$, and $v_{\rm eff}^{\rm E}$ are the electron energy, the wave number of the component of the wave function parallel to the surface, and the total effective potential of the bare electrodes, respectively. The difference between the chemical potentials of the two electrodes can be easily taken into account by adjusting the energies of the highest occupied states (HOSs) of the two groups: the energy of the HOS in $\Psi_{\rightarrow}^{\rm EO}(\boldsymbol{r})$ is set equal to the chemical potential of the left electrode, and that in $\Psi_{\leftarrow}^{\rm EO}(\boldsymbol{r})$ is set equal to the chemical potential of the right electrode.

It is worth mentioning that if one focuses on cases where an infinitesimal bias voltage is applied, the conductance of the system, g, can be evaluated using the following equation [7]:

$$g = \frac{2e^2}{h} {\rm Tr}\left(\Gamma_{\rm L}\, G^{\rm O} \Gamma_{\rm R}\, G^{\rm O}\right). \tag{10.4}$$

Here, $G^{\rm O}$ is the Green's function of the nanoscale objects, which includes the self-energy terms due to the infinite electrodes, and $\Gamma_{\rm L}$ and $\Gamma_{\rm R}$ describe the coupling of the objects to the left and right electrodes, respectively. The above expression can be incorporated into both density functional [8] and molecular-orbital [9] calculations.

Methods Based on Connection of Wave Functions. The basic idea of this approach is simple. This is that the whole system is divided into several regions, the wave functions in these regions are calculated, and then the wave function for the whole system is obtained by connecting the wave functions in the various regions so as to satisfy the boundary conditions, namely the continuity of the wave functions and their derivatives. In the case of nanomolecular systems, it is usually easy to obtain the wave functions in the regions deep inside the semi-infinite electrodes. On the other hand, to obtain the wave functions in the middle region between the electrodes is not necessarily easy, and methods to do this need to be developed. Several methods have been proposed for this purpose, which we describe in the following, taking the one-dimensional case as an example for simplicity.

In the simplest method, the middle region is further divided into thin layers. In the Lth layer, the wave function at a given energy can be expressed as $A_L\psi_L^{\rightarrow}(x) + B_L\psi_L^{\leftarrow}(x)$, where $\psi_L^{\rightarrow}(x)$ is a solution of the Schrödinger (or the Kohn–Sham) equation moving from left to right (or decaying toward the left), and $\psi_L^{\leftarrow}(x)$ is defined similarly. The coefficients A_L and B_L are related to A_{L+1} and B_{L+1} as follows:

$$\begin{pmatrix} A_{L+1} \\ B_{L+1} \end{pmatrix} = T_L \begin{pmatrix} A_L \\ B_L \end{pmatrix}. \tag{10.5}$$

Here, T_L is the transfer matrix, elements of which are determined from the continuity of the wave functions and their derivatives at the boundary between the Lth and $(L+1)$-th layers. Once all the T_Ls have been determined and the coefficients for one layer have been given, the coefficients for all the remaining layers can be obtained. For example, if the coefficients for the first layer, A_1 and B_1, are provided, those for arbitrary $L > 0$ can be obtained as

$$\begin{pmatrix} A_L \\ B_L \end{pmatrix} = T_{L-1} \cdots T_2 T_1 \begin{pmatrix} A_1 \\ B_1 \end{pmatrix} . \tag{10.6}$$

Solutions of the Kohn–Sham equation in each layer can easily be obtained by assuming that the potential is constant in the layer: this assumption is reasonable if the layer is sufficiently thin. Alternatively, one may approximate the kinetic-energy term (the second-order derivative of the wave function) in the Kohn–Sham equation in terms of differences, utilizing the Noumerov method [10], for example. In this case, the wave functions for the whole system can be obtained in a similar way, by considering the values of wave functions at discretized mesh points instead of the coefficients A and B.

Unfortunately, it is known that the above approach often involves numerical instability in practical computations. One way to overcome this difficulty is to consider not the coefficients themselves, but their ratio between neighboring layers. Since there are two independent wave functions at a given energy in a one-dimensional case, we can define a coefficient matrix C_L as

$$C_L = \begin{pmatrix} A_L^1 & A_L^2 \\ B_L^1 & B_L^2 \end{pmatrix} . \tag{10.7}$$

Then a ratio matrix R_L can be defined as

$$R_L = C_{L+1} C_L^{-1} . \tag{10.8}$$

From (10.5) and (10.8), we can see that the ratio matrices satisfy

$$R_L = T R_{L-1}^{-1} . \tag{10.9}$$

This approach, called the recursion transfer matrix method, is known to be much more stable numerically than the approach using (10.6), and has been successfully applied to the electrical properties of nanoscale systems under the influence of electric fields and currents [11].

Another method to avoid numerical instability has been developed by our group recently [12]. In contrast to the above approaches, where the wave functions or coefficients are determined layer by layer, the wave functions in the middle region are solved for simultaneously in our method. This is achieved by eliminating the unknown transmission and reflection coefficients, t and r, from the expressions for the wave functions deep inside the electrodes (see (10.2) and (10.3)), utilizing the continuity conditions on the wave functions and their derivatives at the boundaries between the electrode regions and the middle region. From this treatment, together with the discretization of the Kohn–Sham equation by approximating the kinetic energies in terms of differences, we can derive equations to solve for the wave functions in the middle region simultaneously [12].

Besides the above methods, another approach has been proposed. This method utilizes particular solutions of differential equations, and has been successfully applied to the electrical properties of carbon nanotubes [13].

Miscellaneous Remarks. Although only an outline of the methods has been given above, their extension to more realistic cases, including three-dimensional atomic arrangements, is straightforward. It is also easy to incorporate them into tight-binding and extended Hückel methods.

It is worth mentioning that the jellium model, where the positive ion cores are replaced by a uniform positive charge density, is often employed to represent semi-infinite electrodes when the above methods are incorporated into ab initio calculations. Although the use of the jellium model is not indispensable, it makes practical calculations much easier.

In connection with this, we comment also on the use of nonlocal pseudopotentials. Since nonlocal pseudopotentials can describe the electronic states more accurately than can local pseudopotentials, it is desirable to combine nonlocal potentials with the methods described in this section. However, this makes practical computational procedures complicated, in particular if realistic atomic arrangements are considered, even in the semi-infinite electrodes. This may be a major reason why most of the studies reported so far have adopted the jellium model for the electrodes. However, a few groups have already succeeded in calculations using both electrodes consisting of bulk crystals, and nonlocal pseudopotentials [13,14].

Finally, we would like to comment on the calculation of forces acting on atoms. Although only a few force calculations have been reported so far for nanoscale systems under the influence of electric fields and currents, this does not imply any methodological difficulty. A formulation to calculate forces under the influence of electric fields and currents is described in [11,15], for example.

10.3 Applications to Simple Nanoscale Systems

In this section, we review the application of the methods described in Sect. 10.2.2 to simple nanoscale systems without organic molecules. Examining such systems is helpful in understanding nanomolecular systems. In Sect. 10.3.1, cases with a single electrode are described, while cases with two electrodes are described in Sect. 10.3.2.

10.3.1 Field Emission from Metal Surfaces

Among the phenomena in one-electrode systems under the influence of strong electric fields, field electron emission is of great technological importance and has been investigated actively for a long time. Although the Fowler–Nordheim theory [16,17] has been widely used to analyze experimental data on field emission, the assumptions adopted in the theory, namely a simple one-dimensional model and the Wentzel–Kramers–Brillouin method, are insufficient for cases involving nanoscale objects. An example of such a case is field emission from a metal surface with nanoscale protrusions ending in single atoms. We have performed density functional calculations for this problem [18] using a method developed in our group [12], because of the following reason. The field emission currents from such systems have multiple peaks in the total energy distribution (TED) spectra, which is in sharp contrast to the case of a macroscopic tip, where there is only a single peak at the Fermi energy E_F in the TED spectra. In the case of field emission from single-atom-terminated protrusions on the W(111) surface, a distinct discrepancy exists between experimental results concerning multiple peaks in the TED spectra: one group obtained multiple peaks [19], while others could not obtain them and claimed that extrinsic effects, such as carburization of the tip apex, were essential for obtaining multiple peaks [20]. Therefore, it is desirable to clarify the origin of this discrepancy.

In our calculation, we took the Al(100) surface as an example. We represented it by four Al layers on a semi-infinite jellium electrode, and considered three pyramidal protrusions, consisting of 5, 14, and 30 Al atoms, respectively, attached to the model. Figure 10.3a shows the calculated TED spectra of the emission current for an applied external field $F = 10$ V/nm. In the case of the 14- and 30-atom protrusions, a secondary peak around -1 eV is clearly seen besides the peak at E_F. This result strongly supports the experimental results reported in [19]. On the other hand, this additional peak is not seen in the case of the 5-atom protrusion. These results suggest that a possible origin of the discrepancy between experiments is different sizes of protrusions.

We have demonstrated that two factors are important to obtaining multiple peaks in TED spectra. The first one is a peak below E_F in the local density of states (LDoS) in front of the apex atom of the protrusion, which

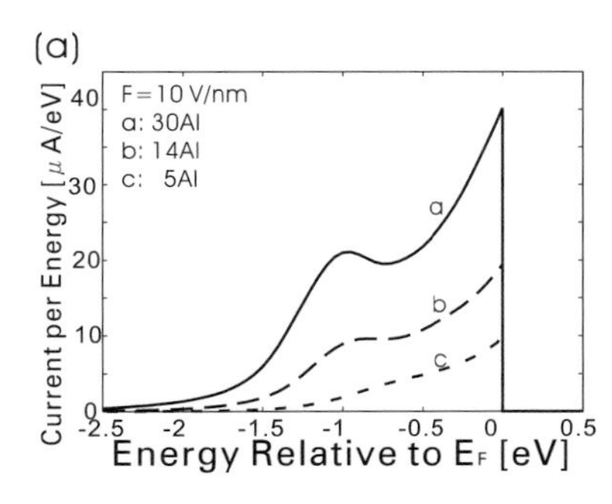

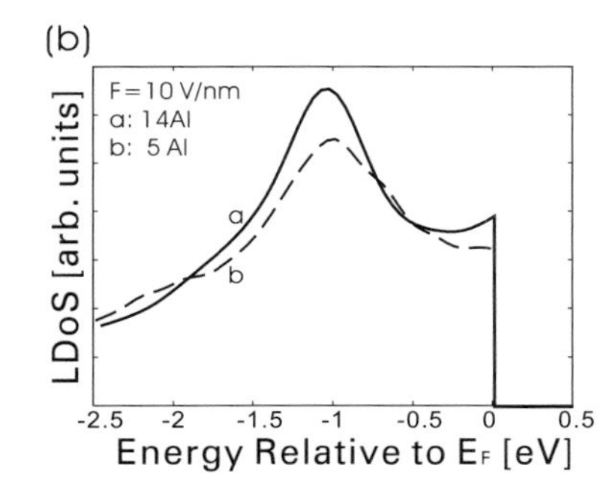

Fig. 10.3. Calculated (**a**) total energy distribution spectra and (**b**) local density of states in front of the apex atom, in the case of $F = 10\,\text{V/nm}$. From ref. [18]

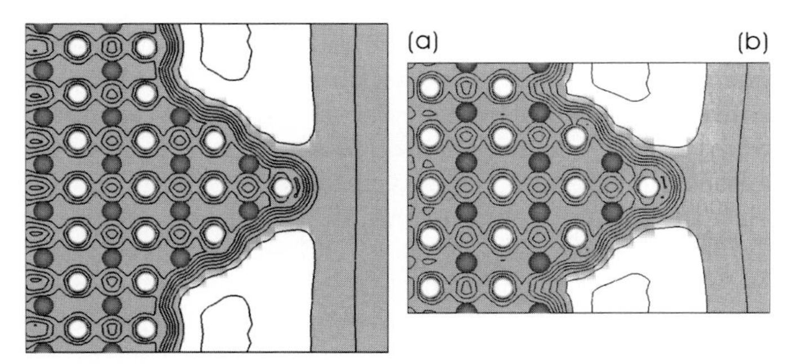

Fig. 10.4. Contour maps of the effective potential V_e for (**a**) a 30-atom and (**b**) a 14-atom protrusions, for $F = 10\,\mathrm{V/nm}$. The interval between the contour lines is 2 eV, and the *dark background* indicates $V_e < E_F$. The *white circles* denote Al ion cores the centers of which are in the plane, and the *dark circles* denote the cores the centers of which are out of the plane. From ref. [18]

is shown in Fig. 10.3b. This peak coincides well with the secondary peak in the TED spectrum, and can be attributed to states localized in front of the apex atom [18]. The peak in the LDoS, however, cannot explain why the secondary peak does not appear in the case of the 5-atom protrusion. The other important factor is the local reduction of the potential barrier for emitted electrons in front of the apex atom, which can be seen in the contour maps of the effective potential shown in Fig. 10.4. It should be emphasized that this barrier reduction depends on the size of the protrusion: the maximum of the potential in front of the apex atom is $-0.87\,\mathrm{eV}$ and $-1.14\,\mathrm{eV}$ for the 14- and 30-atom protrusions, respectively, in the case of $F = 10\,\mathrm{V/nm}$.

From the above two factors, together with the energy dependence of the tunneling probability from the metal to the vacuum, we can understand the behavior of the calculated TED spectra [18]. In this way, our method has succeeded in clarifying the field emission from metal surfaces with nanoscale protrusions ending in single atoms.

10.3.2 Atomic Wires Between Metal Electrodes

Among nanoscale systems with two electrodes, point contacts and atomic wires between two metal electrodes have been actively studied, stimulated by interesting experimental results such as conductance quantization [21]. One of the important issues concerning these systems is their atomic structure. For example, the origin of the unusually long distances between atoms observed in transmission electron microscopy experiments [22] has not been clarified yet, and thus has been studied extensively using various theoretical approaches. However, the methods described in Sect. 10.2 have seldom been applied to this issue, because these methods requires heavy computation to optimize

Table 10.1. Comparison of calculated values of the conductance of an Al_3 chain between jellium electrodes for an infinitesimal bias voltage, together with the types of pseudopotential and methods adopted in the respective calculations

Reference	[6]	[24]	[25]	[23]
Conductance (in $2e^2/h$)	1.6	1.6	1.9	2.0
Type of pseudopotential	NL[a]	NL	L[b]	L
Type of method	G[c]	G	W[d]	W

[a] Nonlocal pseudopotential.
[b] Local pseudopotential.
[c] Green's function.
[d] Connection of wave functions.

the atomic structures of systems having many atoms. Therefore, we skip this issue here.

Another interesting issue is the electrical properties, in particular the conductance, of atomic wires between electrodes. Many calculations related this topic using the methods described in Sect. 10.2 have already been performed, including our own work [23]. In particular, for the case of an Al_3 chain between jellium electrodes, results calculated using different methods have been reported, and so we can compare these results. In Table 10.1, calculated values of the conductance of an Al_3 chain between jellium electrodes for an infinitesimal bias voltage, obtained by different groups and/or methods, are compared. The agreement among the values is reasonabe. However, one may notice a difference between the first two methods [6,24] and the last two [25,23]. Possible origins of this difference are the following. First, the fist two methods adopted a local pseudopotential, while the last two adopted a nonlocal one. Second, the first two methods are based on the connection of wave functions, while the last two use Green's functions. In principle, the two types of methods should give the same results if other computational conditions are the same. On the other hand, the choice of the type of potential may cause differences in the electronic structures. Therefore, we speculate that most of the above discrepancy can be attributed to the difference in the type of pseudopotential used.

Various other interesting findings have been obtained from calculations using the methods described in Sect. 10.2. For example, first, the conductance of a single Na atom between jellium electrodes was shown to be much smaller than that of an Na_2 wire, and this is explained by the difference in the local density of states near the Fermi energy between the two cases [26]. Second, the conductance of bent wires was shown to be different from that of the corresponding straight wires [27,28]. Third, the conductance of a carbon atomic wire was found to oscillate with an increase in its length [29]. Further, the behavior of the conductance of two carbon atomic wires connected in

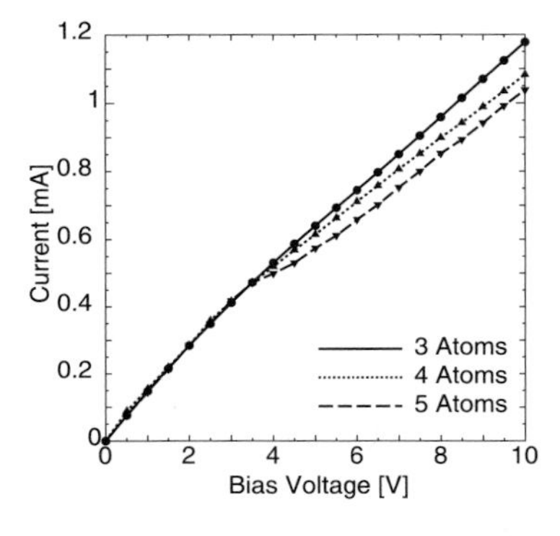

Fig. 10.5. Calculated current–voltage characteristics of Al_3, Al_4, and Al_5 atomic wires between jellium electrodes. From ref. [23]

parallel was found to be distinctly different from that of a single wire, showing a complicated dependence of the conductance on the distance between the two wires [30].

Recently, we have also examined the electrical properties of atomic wires, and have obtained interesting features besides the above [23]. Figure 10.5 shows the calculated current (I)–voltage (V) characteristics of Al_3, Al_4, and Al_5 wires between jellium electrodes. We can see that the I–V curves for the three cases roughly coincide in the range between 0 and 3 V, though an oscillatory behavior of the conductance at ~ 0 V similar to that reported for C [29] and Na [31] wires is also seen in our calculation. Interestingly, distinct differences in the I–V curves are seen among the three wires for $V > 3$ V: with an increment of the bias voltage, the current increases most rapidly in the Al_3 wire, and least in the Al_5.

Differences among the wires are also seen in the potential profiles along the wires shown in Fig. 10.6. In the case of the Al_3 wire with an applied bias voltage of 6 V, the values of the potential at the positions of the atoms change nearly linearly. In contrast, in the case of the Al_5 wire with the same

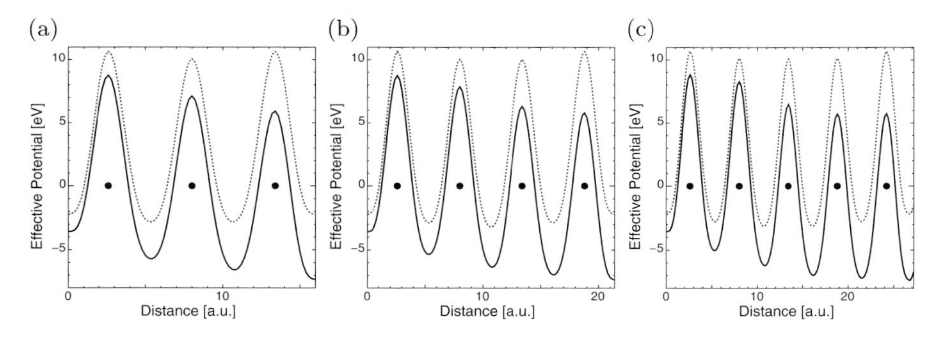

Fig. 10.6. Calculated potential profiles of (**a**) Al_3, (**b**) Al_4, and (**c**) Al_5 wires along the wire. The *solid* and *dotted lines* denote profiles for applied bias voltages of 0 and 6 V, respectively

bias voltage, the change is distinctly nonlinear: the potential change between the second and third atoms from the left in Fig. 10.6 is clearly much larger than that in other regions. In the case of the Al_4 wire, a behavior somewhat similar to that of the Al_5 wire is seen. Although detailed analysis of the above differences among the Al wires is still in progress, we speculate that the behavior of the Al_4 and Al_5 wires can be understood in a similar way to that of semiconductor superlattices [32]. That is, as the bias voltage increases, it becomes energetically unfavorable to maintain band conduction. As a result, a region where the potential change is concentrated is formed at a certain bias voltage. Formation of such a region weakens the potential change in other regions, and thus the band conduction is maintained except in the region where the potential change is concentrated. It is worth mentioning that a potential profile similar to that of the Al_4 and Al_5 wires was seen in a recent self-consistent tight-binding calculation for an Au point contact including the effects of electric fields and currents [33].

10.4 Approaches to Nanomolecular Systems

Compared with the cases of simple nanostructures reviewed in Sect. 10.3, nanomolecular systems consisting of single molecules and electrodes are expected to show a much greater variety of properties, because there are numerous species of molecules, and single molecules themselves can have complex inner structures. Many theoretical studies to clarify the properties of nanomolecular systems have been reported, as described in a recent review on molecular electronics [34]. In fact, the pioneering work in molecular electronics by Aviram and Ratner [35] was also a theoretical study, concerning a single-molecule rectifier.

In this section, we review recent theoretical studies of the electrical properties of nanomolecular systems, focusing on those using the methods described in Sect. 10.2. Such theoretical studies can be categorized into two groups, empirical and self-consistent approaches. The former and the latter are reviewed in Sects. 10.4.1 and 10.4.2, respectively.

10.4.1 Empirical Approaches

Because self-consistent methods including semi-infinite electrodes, electric fields, and currents have been developed only recently, many of the studies reported so far have adopted empirical approaches. Several interesting findings have been obtained from calculations using these approaches as follows. First, from calculations based on an extended Hückel Hamiltonian, it has been clarified how the conductance of nanomolecular systems depends on the molecular species, the number of units in the oligomers, and the geometry of the molecule–electrode interface [36]. Using the same method, the case where two molecules form a bridge between electrodes has also been examined, and it was found that the sum rules of electric resistance valid for macroscopic circuits are not applicable to cases where two molecules are very close to each other [37]. Next, on the basis of a non-self-consistent tight-binding calculation, the existence of huge loop currents inside a molecule was predicted [38,39]: when a molecule containing π-conjugate rings forms a bridge between two electrodes such that the arrangement has no axial symmetry along the direction perpendicular to the electrode surfaces, the magnitude of the loop current inside the molecule becomes much larger than the total current between the electrodes. This is a novel phenomenon peculiar to nanomolecular systems, and its experimental verification is highly desirable.

In the above examples, the effects of the bias voltage between the electrodes were not taken into account. One way to include these effects in the empirical approaches is to adjust the diagonal terms of the Hamiltonian matrix depending on the atomic positions [40]. However, in this approach, the profile of the electrostatic potential when a bias voltages is applied needs to be known (or assumed) before calculation. Since this potential profile depends strongly on the system and sometimes shows complex behavior, as seen in Fig. 10.6, the validity of results obtained assuming a simple potential profile needs to be confirmed using more reliable methods, that is, self-consistent methods where the electron distribution and potential profile are determined self-consistently.

10.4.2 Self-Consistent Approaches

Self-consistent approaches can be further divided into two approaches, namely semiempirical and ab initio ones. Both approaches have already been applied to nanomolecular systems.

In the case of the semiempirical approaches, not only two-electrode systems but also three-electrode ones have been examined, including the effects of finite bias voltage [41]. The variation of current–voltage characteristics when the voltage applied to the gate (the third electrode) is varied has been clarified. Since three-terminal electronic devices are used more widely than two-terminal ones, methods that are able to include the third electrode, such as the one used in [41], are of great use. It should be noted, however, that the reliability of semiempirical calculations may be insufficient in some cases: because the parameters used in the semiempirical methods are usually determined so as to reproduce the atomic and electronic structures of various molecules *without* electric fields and currents, the reliability of the methods in cases containing electric fields and currents is not guaranteed.

Because of this, ab initio methods, which are considered to be more reliable than semiempirical ones, are desirable. An interesting example of such an approache is a calculation of the current–voltage characteristics of a benzene dithiolate molecule between two jellium electrodes [42]. The calculated results agreed with the experimental results [43] qualitatively, but the calculated conductance was two orders of magnitude larger than the observed value. Interestingly, it was also shown that the quantitative agreement could be much improved by inserting an Au atom between the molecule and the electrode [42]. This suggests that the effects of interface geometry on the conductance are very large. Further, the density of states of the system was shown to change significantly when a bias voltage was applied. This indicates the limitation of the empirical approaches, where such a change in the density of states is neglected. In addition, the same group has examined cases where a gate electrode is put near the molecule, though the electron transfer between the molecule and the gate is neglected [44]. Very recently, this group has also examined the current-induced forces and distortions in a nanomolecular system [45].

Although the calculation appeared to have fully explained the experimental data [43], another interpretation of the experimental data has been proposed very recently, on the basis of semiempirical calculations [46]. In contrast to the previous calculation [42], where a molecule is assumed to be attached to two electrodes, each molecule is attached to only one electrode in the recently proposed model [46]. Further studies are necessary to judge which model describes the experiments correctly.

As for ab initio approaches to other nanomolecular systems, the case where a hydrogen atom in benzene dithiolate is substituted by a nitro group was examined, and in this case the current–voltage characteristics were shown to be sensitive to the rotation of the nitro group [47]. Several other di-substituted benzenes, as well as pyrazine and biphenyl, between two jellium electrodes have also been examined recently [48]. In addition, calculations for a C_{20} molecule between Al and Au electrodes have been reported [14]. Considering the high reliability of ab initio approaches, the number of calculations for nanomolecular systems using these approaches is expected to increase rapidly.

10.5 Concluding Remarks

In this chapter, we have described some theoretical approaches used to examine the electrical properties of nanomolecular systems, including the effects of semi-infinite electrodes, electric fields, and currents. We have also reviewed several examples, where these approaches were applied to simple nanoscale systems, nanoscale protrusions on a metal surface, atomic wires between two electrodes, and nanomolecular systems. Most of these approaches have been developed only recently, and thus calculations using these approaches have been much fewer than those using the conventional methods. However, these new approaches are expected to become increasingly important.

It should be noted that though these approaches have succeeded in clarifying interesting features of the properties of various nanoscale systems, including nanomolecule systems, there remain many problems for further theoretical studies. First, the effects of inelastic scattering are not taken into account in the methods described in Sect. 10.2. Inclusion of only elastic scatterings may be valid in cases where the length of the molecules is shorter than the mean free path, such as the case of benzene dithiolate between metal electrodes. In other cases, however, it is desirable to include the effects of inelastic scattering. A formalism to incorporate these effects into the methods described in Sect. 10.2 has already been proposed [49]. Next, the methods in Sect. 10.2 are insufficient in cases where the electron–phonon or electron–electron interaction is very strong. The former interaction is important in molecules where polaronic electronic conduction is expected, while the latter may be important in molecules that have strong one-dimensional character and are thus expected to show the characteristics of a Tomonaga–Luttinger liquid. To develop methods appropriate for such cases is a challenging problem. Finally, the time evolution of electronic states is also neglected in the methods in Sect. 10.2, which treat only steady states. This issue might be important, for example, in predicting the operation of nanomolecular devices, and thus is an interesting future problem.

Acknowledgments

The author thanks Prof. K. Watanabe, Prof. H. Fujita, Dr. Y. Nakamura, Dr. N. Sasaki, Mr. Y. Gohda, Mr. S. Furuya, Mr. N. Nakaoka, and Dr. K. Tada for collaboration and useful discussions during our work presented in this chapter. Mr. Y. Gohda, Mr. S. Furuya, and Mr. N. Nakaoka are acknowledged also for providing figures. The work was supported by the Core Research for Evolutional Science and Technology (CREST) program of the Japan Science and Technology (JST) Corporation.

References

1. N. Nakaoka, K. Tada, S. Watanabe, H. Fujita, K. Watanabe: Phys. Rev. Lett. **86**, 540 (2001)
2. J.R. Chelikowsky, N. Troullier, Y. Saad: Phys. Rev. Lett. **72**, 1240 (1994)
3. T. Hoshi, M. Arai, T. Fujiwara: Phys. Rev. B **52**, R5459 (1995)
4. J. Tersoff, D.R. Hamann: Phys. Rev. B **31**, 805 (1985)
5. M. Tsukada, N. Shima: J. Phys. Soc. Jpn. **56**, 2875 (1987)
6. N.D. Lang: Phys. Rev. B **52**, 5335 (1995)
7. S. Datta: *Electronic Transport in Mesoscopic Systems* (Cambridge University Press, Cambridge 1995)
8. M.B. Nardelli, J.-L. Fattebert, J. Bernholc: Phys. Rev. B **64**, 245423 (2001)
9. S.N. Yaliraki, A.E. Roitberg, C. Gonzales, V. Mujica, M.A. Ratner: J. Chem. Phys. **111**, 6997 (1999)
10. B.V. Noumerov: Mon. Not. R. Astron. Soc. **84**, 592 (1924)
11. K. Hirose, M. Tsukada: Phys. Rev. B **51**, 5278 (1995)
12. Y. Gohda, Y. Nakamura, K. Watanabe, S. Watanabe: Phys. Rev. Lett. **85**, 1750 (2000)

13. H. Choi, J. Ihm: Phys. Rev. B **59**, 2267 (1999)
14. C. Roland, B. Larade, J. Taylor, H. Guo: Phys. Rev. B **65**, 041401 (2002)
15. M. Di Ventra, S.T. Pantelides: Phys. Rev. B **61**, 16207 (2000)
16. R.H. Fowler, L.W. Nordheim: Proc. R. Soc. London A **119**, 173 (1928)
17. E.L. Murphy, R.H. Good: Phys. Rev. **102**, 1464 (1956)
18. Y. Gohda, S. Watanabe: Phys. Rev. Lett. **87**, 177601 (2001); Surf. Sci. **516**, 265 (2002)
19. V.T. Binh, S.T. Purcell, N. Garcia, J. Doglioni: Phys. Rev. Lett. **69**, 2527 (1992)
20. M.L. Yu, N.D. Lang, B.W. Hussey, T.H.P. Chang, W.A. Mackie: Phys. Rev. Lett. **77**, 1636 (1996)
21. J.I. Pascual, J. Méndez, J. Gómez-Herrero, A.M. Baró, N. García, V.T. Binh: Phys. Rev. Lett. **71**, 1852 (1993)
22. H. Ohnishi, Y. Kondo, K. Takayanagi: Nature **395**, 780 (1998)
23. S. Furuya, Y. Gohda, N. Sasaki, S. Watanabe: Jpn. J. Appl. Phys. **41**, L989 (2002)
24. N. Kobayashi, M. Aono, M. Tsukada: Phys. Rev. B **64**, 121402(R) (2001)
25. N. Kobayashi, M. Brandbyge, M. Tsukada: Jpn. J. Appl. Phys. **38**, 336 (1999)
26. N.D. Lang: Phys. Rev. Lett. **79**, 1357 (1997)
27. N. Kobayashi, M. Brandbyge, M. Tsukada: Phys. Rev. B **62**, 8430 (2000)
28. N.D. Lang, P. Avouris: Phys. Rev. Lett. **84**, 358 (2000)
29. N.D. Lang, P. Avouris: Phys. Rev. Lett. **81**, 3515 (1999)
30. N.D. Lang, P. Avouris: Phys. Rev. B **62**, 7325 (2000)
31. H.-S. Sim, H.-W. Lee, K.J. Chang: Phys. Rev. Lett. **87**, 096803 (2001)
32. L. Esaki, L.L. Chang: Phys. Rev. Lett. **33**, 495 (1974)
33. M. Brandbyge, N. Kobayashi, M. Tsukada: Phys. Rev. B **60**, 17064 (1999)
34. C. Joachim, J.K. Gimzewski, A. Aviram: Nature **408**, 541 (2000)
35. A. Aviram, M. Ratner: Chem. Phys. Lett. **29**, 277 (1974)
36. M. Magoga, C. Joachim: Phys. Rev. B **56**, 4722 (1997)
37. M. Magoga, C. Joachim: Phys. Rev. B **59**, 16011 (1999)
38. S. Nakanishi, M. Tsukada: Jpn. J. Appl. Phys. **37**, L1400 (1998)
39. S. Nakanishi, M. Tsukada: Phys. Rev. Lett. **87**, 126801 (2001)
40. V. Mujica, M. Kemp, A. Roitberg, M. Ratner: J. Chem. Phys. **104**, 7296 (1996)
41. E.G. Emberly, G. Kirczenow: Phys. Rev. B **62**, 10451 (2000)
42. M. Di Ventra, S.T. Panterides, N.D. Lang: Phys. Rev. Lett. **84**, 979 (2000)
43. M.A. Reed, C. Zhou, C.J. Muller, T.P. Burgin, J.M. Tour: Science **278**, 252 (1997)
44. M. Di Ventra, S.T. Panterides, N.D. Lang: Appl. Phys. Lett. **76**, 3448 (2000)
45. M. Di Ventra, S.T. Panterides, N.D. Lang: Phys. Rev. Lett. **88**, 046801 (2002)
46. E.G. Emberly, G. Kirczenow: Phys. Rev. B **64**, 235412 (2001)
47. M. Di Ventra, S.T. Panterides, N.D. Lang: Phys. Rev. Lett. **86**, 288 (2001)
48. N.D. Lang, P. Avouris: Phys. Rev. B **64**, 125323 (2001)
49. E.G. Emberly, G. Kirczenow: Phys. Rev. B **61**, 5740 (2000)

11 Nanodevices for Quantum Computing Using Photons

Shigeki Takeuchi

Summary. The amount of research aiming to apply essential features of quantum theory such as the uncertainty principle, the quantum interference of wave functions, and quantum entanglement to information communication and information processing has been remarkable recently. For example, quantum cryptography, which completely excludes the possibility of eavesdropping, becomes possible when the uncertainty principle is applied. Also, the quantum computer has been proposed. By fully applying quantum superposition states and quantum entanglement to computation, a computation that would take billions of years using present supercomputers might be performed in several minutes. The research into quantum computers is still in its early stages, but for quantum cryptography, verification experiments over tens of kilometers have already been performed in various places and research aimed at practical use is proceeding. In this chapter, devices of all kinds which are needed in quantum cryptography and quantum computation are called quantum information devices.

Nanotechnology and quantum information devices are related very closely. For example, the discrete quantum states which are used to store the information in quantum computation can be observed only in nanostructures where the numbers of electrons and atoms are very small, and cannot be seen in macroscopic materials. Also, during a quantum computing process, it is required that the quantum superposition state should never be broken. The time during which the superposition state is maintained is called the decoherence time. To prevent decoherence, the interactions with the outside which cannot be controlled must be blocked off. For this purpose, various nanotechnologies are required. In this chapter, the general concepts of quantum communication and quantum information processing are explained first, and then we describe the devices which are needed for quantum cryptography and quantum computers to be realized, while aiming specifically to deal with their relation to nanotechnology. In the last part of the chapter, we shall focus on nanodevices for quantum computing using photons.

11.1 Quantum Communication and Quantum Information Processing

When we hear the words "the quantum theory", we may have an image that it has no relation to daily life or simply that it is very difficult. Indeed, in the quantum theory, strange features such as quantum interference of wave functions, the uncertainty principle, and quantum entanglement play essential roles. Since the quantum theory was discovered in the 1920s, arguments have continued concerning the truth and the interpretation of these features. The existence of these features, however, has been verified by various experiments. The devices such as the transistor and the laser, which are indispensable for our daily life, were invented on the basis of the quantum theory.

Around the end of the 1960s, by utilizing those strange features of quantum theory, an idea to produce bills that can never be copied was presented. At that time, however, the experiments to verify the quantum theory had not been performed to a sufficient extent and the idea was seen as a castle in the air. Later, Bennett and Brassard sublimated the idea into "quantum cryptography" (the BB84 protocol), which uses the collapse of wave functions and the uncertainty principle, in 1984 [1]. By using this idea, a secret key with any amount of bits can be shared without an eavesdropper peeking. When a secret key with a length that is the same as that of the sentence to be encrypted is prepared in this way, the deciphering of the encrypted communication becomes impossible.

Next, quantum computation was proposed in 1985 [2]. This is the idea that makes large-scale computation possible by applying the principle of quantum superposition and quantum entanglement. At first, the importance of the idea was not sufficiently recognized. However, when Shor's factorization algorithm [3] was invented in 1994, the idea attracted a great deal of attention. It is known that the computing time for a factorization increases as an exponential function of the number of input digits. It has also been said that it would take hundreds of millions of years to factorize a 200-digit number even if one used the highest-speed computer currently available. This fact guarantees the safety of the RSA cryptography that is widely used in the Internet at present. However, Shor showed that the computation time required scales linearly with the number of the digits when quantum computation is used. If we assume a quantum computer with a 100 MHz clock, the computation time is decreased to a few minutes for a 200-digit factorization. This ultrafast computational ability is the most important characteristic of quantum computation.

11.2 Quantum Cryptography

11.2.1 The Mechanism of Quantum Cryptography

First, let us think about eavesdropping in present-day communication systems. For example, in an existing communication system that uses light pulses for transmission through optical fibers, the light can be easily detected outside by just bending the optical fiber a little. If the quantity of the leakage is small, it is difficult to detect the eavesdropping. Therefore, the data is encrypted on the assumption that someone might eavesdrop on any data. Usually, relatively short secret keys are used for the purpose of quick encryption. Such encryption systems using short keys, however, may be deciphered. According to information theory, it becomes impossible to decipher a message only when a random-number table with a length that is the same as that of the original data line is used for encryption. The question arises, is it possible for such a random-number table to be shared without being known by anyone else?

When one uses an extremely feeble light pulse, each of the photons in the pulse can be detected with a photon detector. Under such a condition, a feature of quantum theory appears which is not seen in a strong light field. This situation is shown in Fig. 11.1.

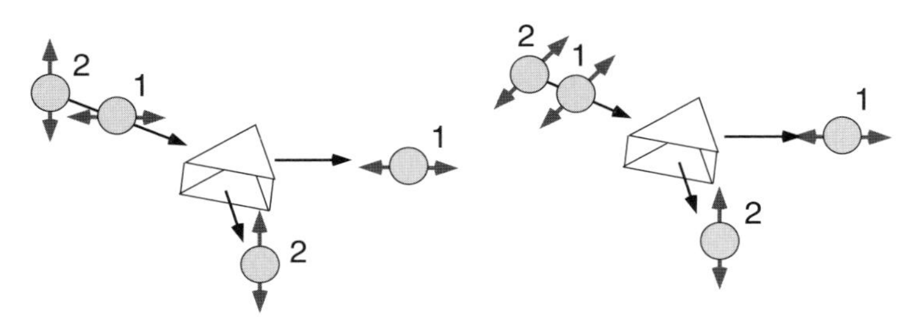

Fig. 11.1. Measurement of polarization of single photons

Let us consider the case (Fig. 11.1, left) where the input photon could have a vertical or horizontal polarization. In this case, vertically polarized photons and horizontally polarized photons can be separated and be detected without making a mistake using a polarizing beam splitter (a special prism). However, when a circularly polarized photon is input into the detection equipment, what happens (Fig. 11.1, right)? At this time, a photon is taken to be either a "vertically polarized photon" or a "horizontally polarized photon" with probability 1/2. In the example of Fig. 11.1 (right), two obliquely polarized photons are incident on the equipment; however, the first photon is read as a vertically polarized photon and the second one is read as a horizontally polarized photon. This is caused by the uncertainty relation between linear polarization and oblique polarization. In other words, one cannot *guess* the (embedded) polarization of a single photon with 100% certainty.

Quantum cryptography uses this mechanism successfully. The purpose of quantum cryptography is to share a random-number table between two separated parties without any eavesdropping. The BB84 protocol [1] is explained in Fig. 11.2. First, the sender and receiver prepare two kinds of code table, "+" and "×", which show the polarization of light corresponding to a certain bit. The sender selects one of the two tables randomly for each of the bits in the random-number table, and transmits a photon which has the polarization shown in the table. Note that information on the selection of the code tables should not be disclosed at this time. In this example, the transmission code "+" is randomly chosen for the head of the random-number table ("1"). Then,

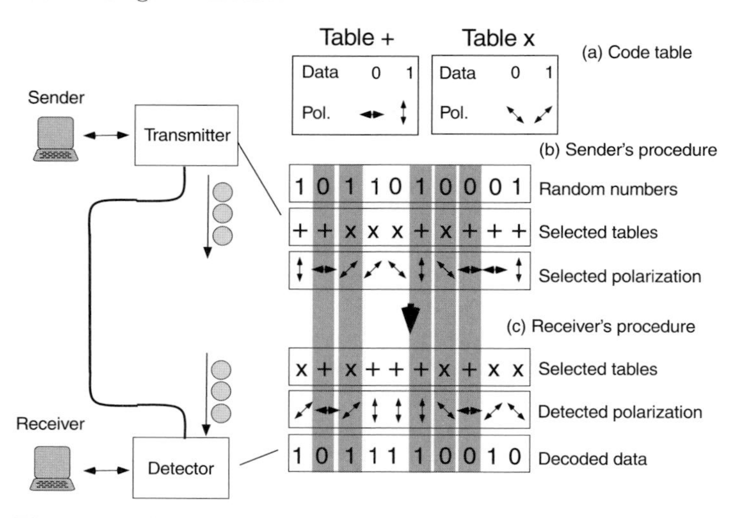

Fig. 11.2. The procedure for the BB84 quantum key distribution protocol

a *vertically polarized photon* is transmitted for the data bit "1" according to the code table "+".

The receiver chooses one of the two code tables randomly in the same way and reads the polarization of light of the photon in the direction that corresponded to the table. When the code tables which were used by the sender and receiver are the same (shaded in Fig. 11.2) by chance, the bit value transmitted by the sender and the bit value decoded by the receiver become the same. On the other hand, when the code tables are different, this process becomes inaccurate.

Let us see an example. The first photon is sent with a vertical polarization because the sender selected the code table "+". Unfortunately, the receiver selected the code table "×" in this case. As a result, the measurement result should be one of the "obliquely polarized states".

When the sender and receiver have completed the above procedure for a sequence of data, they have to check whether they have used the same code table or not for each of the bits. At this stage, they may exchange their records using usual communication methods. After the check, they can share a random key by gathering together the bits that were sent and received using the same code table. In the example shown in the Fig. 11.2, a secret key "01100" is shared.

Next, the way to find an eavesdropper will be explained. For the eavesdropper, it is not possible to know the polarization of photons in the transmission line. In order to obtain information from a photon, the eavesdropper has to detect the photon according to the code table used for the decoding. The probability that the eavesdropper finds the correct one is just 50%. If the photon has disappeared, it is easy for the receiver to notice the bugging. In order to escape detection, the eavesdropper must resend the photon. However, when the code table is wrong, the eavesdropper will send the photon with the wrong polarization. In such a case, the shared data that would normally agree fully should differ. Therefore, the thing that the sender and receiver must do in order to find an eavesdropper is to compare the shared data occasionally. The data may have been subjected to bugging when some discrepancies are found in the shared secret key. The points mentioned above are the mechanism of quantum cryptography.

11.2.2 Devices for Quantum Cryptography

It may be doubted whether it is possible to send and receive such photons one by one, and control the polarization of each photon. However, this is possible by use of existing optical communication technology. In present research, laser light is simply attenuated by neutral-density filters so as to give an average number of photons in a pulse of about 0.1. The polarization of the photons can be rotated by use of existing electrooptic devices. Also, for long-range transmission, it is possible to use an optical fiber of the usual kind for the transmission line. Zbinden et al. at Geneva University succeeded in 1998 in generating a key at 210 bits per second using the optical fiber 23 km in length laid under Lake Geneva [4]. In Japan, in 2000, in a cooperative research project of Mitsubishi Electric and ourselves, we have succeeded in generating a key of 1 kbit per second with a transmission distance of 200 m [5].

The increase in the distance and the speeding up of quantum cryptography are future issues. For this purpose, quantum information devices such as the "quantum phase gate", which is necessary to relay the quantum information over long distances, and "single-photon sources", which generate the photons one by one, will be needed. These are also key devices for quantum computers using photons, and are described in Sect. 11.4.

11.3 Quantum Computation

11.3.1 The Mechanism of Quantum Computation

Quantum computation is an idea that makes incredibly massive parallel computation possible using quantum superposition states and quantum entanglement [6]. In present computers, the logic is based on "bits", which take values of 0 or 1. A logic circuit is made with logic gates, which transform the value of bits. On the other hand, quantum computation uses quantum logic gates acting on *quantum bits*, or *qubits*, which can be superposition states of 0 and 1. These superposition states can be described using a parameter θ and a phase parameter α. Using those parameters, a quantum bit can be visualized as a vector from the origin to a point on the surface of a sphere with radius 1, as shown in Fig. 11.3. Here, the symbol $|\rangle$ is called a "ket", indicating that it is a quantum state (or, in other words, a wave function).

$$|a\rangle = \cos\frac{\theta}{2}|0\rangle + \exp(i\alpha)\,\sin\frac{\theta}{2}|1\rangle$$

Fig. 11.3. A visualization of a quantum bits

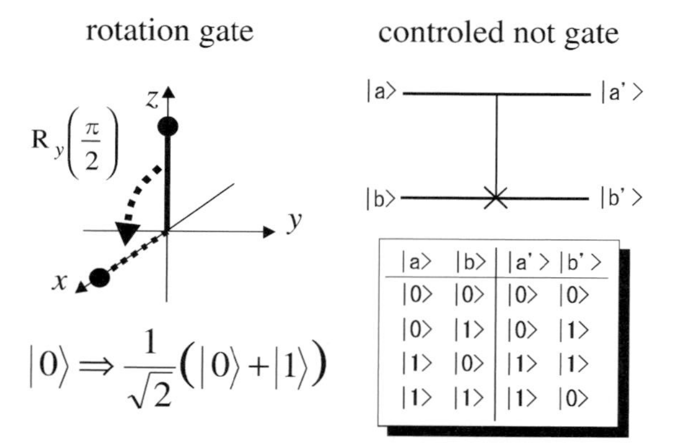

|a>	|b>	|a' >	|b' >
|0>	|0>	|0>	|0>
|0>	|1>	|0>	|1>
|1>	|0>	|1>	|1>
|1>	|1>	|1>	|0>

$$R_y\left(\frac{\pi}{2}\right) \qquad |0\rangle \Rightarrow \frac{1}{\sqrt{2}}\left(|0\rangle + |1\rangle\right)$$

Fig. 11.4. Basic quantum gates: a rotation gate and a controlled-NOT gate

If two kinds of "basic quantum gates" can be realized, a quantum computer will become possible (Fig. 11.4). One of the gates is called a "rotation gate", and acts on a single quantum bit so as to "rotate" the vector corresponding to the bit. The other basic quantum gate is called the "controlled-NOT gate", and acts on two quantum bits, namely a signal bit and a target bit. This acts on the target bit as a NOT gate only when the signal bit is 1.

As a concrete example, let us look at a Fourier transform circuit to understand how a quantum computer works. An example of a quantum Fourier transform circuit is shown in Fig. 11.5. If the binary-number representation of $|a\rangle$ is $|a1\rangle|a0\rangle$ and similarly for $|b\rangle$, $|b\rangle$ becomes the discrete Fourier trans-

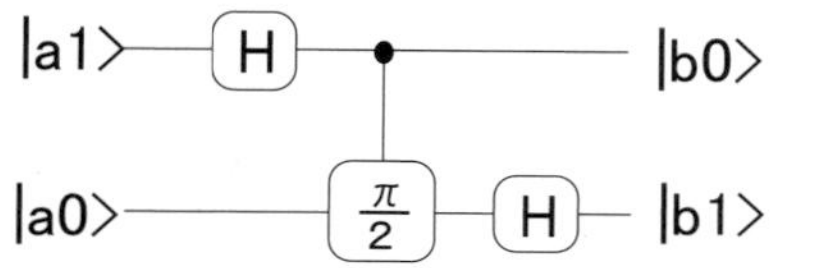

Fig. 11.5. A quantum circuit for two-bit Fourier transformation

form of $|a\rangle$. In this example, a discrete Fourier transformation of a two-bit number is performed by three gate operations. In a similar way, the discrete Fourier transformation of an N-bit number can be executed with $N(N+1)/2$ operations. When we perform a Fourier transformation of an N-bit number straightforwardly using classical methods, it needs 2^{2N} steps. Even when we adopt the fast Fourier transform (FFT) method, it takes $N \times 2^N$ steps. Now you can see how overwhelmingly fast a quantum computer can solve a typical problem.

Shor discovered an algorithm which solves the problem of factorization and a discrete logarithm problem at high speed, using this quantum Fourier transform successfully [3]. After that, Grover discovered an algorithm which performs a database search at high speed [7]. To find a specific date item in a database in which N data items are stored at random, N trials are required using the conventional way. However, when quantum computation is used, only $\sqrt{N}$ steps are required. For example, when 1,000,000 steps are required for a conventional computer, only 1000 steps are required for a quantum computer.

11.3.2 Towards the Realization of Quantum Computers

To realize such quantum computers, the basic quantum gates mentioned above have to be realized for a specific kind of quantum bit. For these quantum bits, various physical quantities can be used, such as nuclear spins and the energy levels of electrons [8].

So far, the algorithms for quantum computation have been verified by experiments using the nuclear spins of molecules [9] and using single photons with linear optics [10]. However, it is believed that the maximum number of qubits is a few tens when we use these methods.

An ion trap quantum computer is a candidate for larger-scale quantum computation. A four-quantum-bit operation has been demonstrated at Los Alamos National Laboratory [11]. Also, various ideas have been proposed for quantum computation using solid-state devices. A well-known scheme was proposed by Kane [12], which utilizes the nuclear spins of phosphorus atoms embedded in a silicon semiconductor. The realization of solid-state quantum computers has also been attracting the attention of Japanese scientists. For example, Nakamura at NEC has performed a "rotation gate" operation on a quantum bit that corresponds to the number of Cooper pairs in a nanoisland [13]. In addition, the possibility of using coupled quantum dots and SQUIDs as quantum bits has been investigated in detail.

11.3.3 A Hard Bottleneck: Decoherence

When will a quantum computer that surpasses present computers be realized? I do not think that such a quantum computer will be realized by 2010, but specifying the time is very difficult. One of the main issues to be overcome is the decoherence of the superposition states. In quantum computation, superposition states must be maintained during computation. However, as with the actual quantum bits, superposition states are gradually broken as time passes. For example, if we want to factorize a 200-digit number, about 10^{10} gate operations must be performed on about 1000 quantum bits [14]. If we assume that the time for a gate operation is somewhat less than a nanosecond, the state must not decohere for at least several seconds.

In ion trap quantum computers, the decoherence time is expected to be long because the ions are trapped in a vacuum, so that the state is not affected by interaction with environment. However, it seems that it will be quite difficult to arrange thousands of ions in a vacuum and control them. On the other hand, solid-state systems seem good candidates for integrating qubits. However, in this case it is difficult to block off the interaction between a quantum bit and the environment. Basic research on the proposed ideas and then building verification experiments on these proposals will be important.

11.4 Nanodevices for Quantum Computing Using Photons

Among the above proposals, quantum computers using single photons are promising candidates. It is very hard to control and detect the state of a single atomic spin. On the other hand, we can use conventional optical devices such as beam splitters and wave plates to control the states of photons. Measurement of photonic states is also possible with available technologies. For example, we have developed a photon-counting system with a quantum efficiency of 88%, which also has an ability to determine the number of incident photons at the same time [15,16].

Recently, a novel idea was proposed by Knill et, al. [17]. These authors showed that quantum computation is possible using only linear optics, counters that measure photon numbers, and single-photon sources. We have also shown that a beam splitter with a reflectance of 1/3 can act as a quantum phase gate when post-selection is allowed [18]. Here, we shall give an overview of single-photon sources and quantum phase gates.

11.4.1 Single-Photon Sources

A single-photon source is a device which produces only one photon in a pulse. As we have already seen, the average number of photons in a pulse can be made equal to one using existing technologies. In this case, however, the states where the photon number is 0 (the condition in which no photon exists in), 2, and so on are also generated, with probabilities given by the Poisson distribution.

Three methods have been proposed as single-photon sources. The first uses a parametric fluorescence pair [19], the second uses luminescence from a controlled electron–hole pair [20], and the third uses luminescence from single molecules and atoms [21]. The two latter methods need especially cutting-edge nanotechnology.

Kim et. al. at Stanford University, realized a turnstile single-photon source [20]. Here, one electron and one hole are injected into a recombination area using the Coulomb blockade effect. The quintessence of nanotechnology was used to create this device: a semiconductor quantum well was etched to a column tens of nanometers in size. Currently, the improvement of the output efficiency is a challenge.

11.4.2 Photonic Quantum Phase Gates

As part of the effort directed toward the superfast all-optical network system, research on optical Kerr devices, which control the phase of a signal light pulse by means of a control light pulse, is proceeding. A photonic quantum phase gate may be the ultimate optical Kerr device. It modulates the phase of the signal photon by means of the condition of the control photon. The pioneering experiment on photonic quantum phase gates was done by Kimble's group at Caltech in 1995 [22]. In this experiment, a cesium atom that could be confined in an optical cavity with a high-Q light resonance was used as the device. When a photon with a particular polarization is incident on the atom, the atom is excited from the ground state. Owing to the high-Q cavity, it becomes possible for the photon to have a strong interaction with the atom. Thus, a kind of absorption saturation occurs after only one photon has been incident. In such a state, a second incident photon can no longer sense the existence of the atom, and the phase shift caused by the atom in the usual case is no longer detectable. Kimble's group observed a phase difference of 14 degrees in their experiment.

This experiment is a true pioneering work. However, there are still many things to be explored. The group used weak coherent light as a light source and did not succeed in the verification of the phase shift using a single-photon state. The other challenge is to obtain a fixed phase shift under all conditions. In the experiment described above, the amount of phase change was not stable, because it depended strongly on the position of the Cs atom in the cavity. The realization of a quantum phase gate will be a breakthrough in the realization of quantum information systems. In the future, solid-state devices may be useful. Nanotechnologies such as microoptical resonators and photonic crystals will play a key role in such devices.

As previously mentioned, we have shown that a beam splitter with a reflectance of 1/3 can act as a quantum phase gate when post-selection is allowed. This will be useful in some particular cases where the detection of the result just after the gate is allowed.

11.5 Conclusion

In this chapter, a review of quantum information technology and the future prospects for quantum computation devices using photons has been given.

Acknowledgments

The author would like to thank Prof. Milburn, Prof. Rubinstein-Dunlop, Dr. White, and Dr. Ralph for their kind hospitality during his stay at the Physics Department of the University of Queensland.

References

1. C.H. Bennett and G. Brassard: in *Proceedings of the IEEE International Conference on Computer, Systems, and Signal Processing, Bangalore, India* (IEEE, New York 1984), p. 175
2. D. Deutsch: Proc. R. Soc. London Ser. A **400**, 97 (1985)
3. P.W. Shor: in *Proceedings of the 35th Annual Symposium on Foundation of Computer Science* (IEEE Computer Society, Los Alamitos, CA 1994), p. 124
4. H. Zbinden, H. Bechmann-Pasquinucci, N. Gisin, G. Ribordy: Appl. Phys. B **67**, 743 (1998)
5. T. Hasegawa, T. Nishioka, H. Ishizuka, J. Abe, K. Shimizu, M. Matsui, S. Takeuchi: IEICE Trans. Fundam., **E85-A** 149 (2002)
6. A. Ekert, R. Jozsa: Rev. Mod. Phys. **68**, 733 (1996)
7. L.K. Grover: Phys. Rev. Lett. **79**, 325 (1997)
8. D.P. Divincenzo: arXiv:quant-ph/0002077
9. L.M.K. Vandersypen, M. Steffen, G. Breyta, C.S. Yannoni, M.H. Sherwood, I.L. Chuang: Nature **414**, 883 (2001)
10. S. Takeuchi: Phys. Rev. A **62**, 032301 (2000); S. Takeuchi: Phys. Rev. A **61**, 052302 (2000)
11. C.A. Sackett, D. Kielpinski, B.E. King, C. Langer, V. Meyer, C.J. Myatt, M. Rowe, Q.A. Turchette, W.M. Itano, D.J. Wineland, C. Monroe: Nature **404**, 256 (2000)
12. B.E. Kane: Nature **393**, 133 (1998)
13. Y. Nakamura, Yu.A. Pashkin, J.S. Tsai: Nature **398**, 768 (1999)
14. R.J. Hughes, D.F. James, E.H. Knill, R. Laflamme, A.G. Petschek: Phys. Rev. Lett. **77**, 3240 (1996)
15. S. Takeuchi, J. Kim, Y. Yamamoto, H.H. Hogue: Appl. Phys. Lett. **74**, 1063 (1999)
16. J. Kim, S. Takeuchi, Y. Yamamoto, H.H. Hogue: Appl. Phys. Lett. **74**, 902 (1999)
17. E. Knill, R. Laflamme, G.J. Milburn: Nature **409**, 46 (2001)
18. H.F. Hofmann, S. Takeuchi: Phys. Rev. A **66** 024308 (2002)
19. S. Takeuchi: in *Proceedings of the 7th International Symposium on Foundations of Quantum Mechanics in the Light of New Technology (ISQM-Tokyo '01)* Edited by Y.A. Ono and K. Fujikawa (World Sceintific Publishing, Singapore, New Jersey 2002), p. 98

20. J. Kim, O. Benson, H. Kan, Y. Yamamoto: Nature **397**, 500 (1999)
21. B. Lounis, W.E. Moerner: Nature **407**, 491 (2000)
22. Q.A. Turchette, C.J. Hood, W. Lange, H. Mabuchi, H.J. Kimble: Phys. Rev. Lett. **75**, 4710 (1995)

Index